Growing Local

Our Sustainable Future

SERIES EDITORS

Charles A. Francis
University of Nebraska–Lincoln

Cornelia Flora
Iowa State University

Tom Lynch
University of Nebraska–Lincoln

Growing Local

**Case Studies on Local
Food Supply Chains**

*Edited by
Robert P. King,
Michael S. Hand,
and Miguel I. Gómez*

UNIVERSITY OF NEBRASKA PRESS | LINCOLN AND LONDON

Library of Congress
Cataloging-in-Publication Data
Growing local: case studies on local food
supply chains / edited by Robert P. King,
Michael S. Hand, and Miguel I. Gómez.
pages   cm.—(Our sustainable future)
Case studies on local food supply chains
Includes bibliographical references and index.
ISBN 978-0-8032-5485-5 (cloth: alk. paper)
ISBN 978-0-8032-5816-7 (epub)
ISBN 978-0-8032-5819-8 (mobi)
ISBN 978-0-8032-5699-6 (pdf)
1. Food supply—United States—Case studies.
2. Local foods—United States—Case studies.
I. King, Robert Philip, 1950– II. Hand, Michael
S. (Michael Stephen), 1977– III. Gómez, Miguel
I. IV. Title: Case studies on local food supply
chains. V. Series: Our sustainable future.
HD9005.G76 2014
338.1'973—dc23
2014024550

CONTENTS

FIGURES

TABLES

ACKNOWLEDGMENTS

The research in this volume was supported through cooperative research funding from the USDA Economic Research Service (ERS) in Washington DC. The authors thank Mary Bohman of ERS for her thoughtful guidance and support of this research project. Numerous workshop and seminar participants also provided helpful comments during the preparation of this book. The views expressed in this volume are the authors' and do not necessarily represent the views of the U.S. Department of Agriculture.

Data collection assistance for market prices and product availability was provided by Jun Lee (Cornell University), Amber Ma (University of California–Davis), Katherine Ralston (ERS), Joe Rodriguez (Oregon State University), and Sarah Swan (University of Minnesota). Lisa Jore and Sue Pohlod, both from the Department of Applied Economics at the University of Minnesota, assisted with the editing and assembly of this book.

Finally, the authors thank the many farmers, vendors, market managers, and business operators who submitted to interviews and gave their time to provide information about their operations. Some are identified where they are quoted, while others requested confidentiality; all were important to this project. This research would not have been possible without their generous contributions.

Growing Local

Part 1

Understanding Local Food Systems
from a Supply Chain Perspective

1 From Farms to Consumers

An Introduction to Supply Chains for Local Foods

Miguel I. Gómez and Michael S. Hand

The term *local foods* conjures vivid and specific images among consumers, food connoisseurs, and scholars. Many people think of the fresh young vegetables and the first ripe strawberries that appear in farmers markets in the spring and the apples and winter squash that herald fall's arrival at the end of the market season. For others, what comes to mind is a roadside farm stand, discovered by accident during a Saturday drive out of town and packed with a variety of straight-from-the-field produce. More and more, the picture of local foods also includes signs in supermarkets identifying certain products as local, and stories from farmers about how their food was produced.

These images are a growing part of how people think about their food when they fill their grocery cart (or canvas bag or farm share box). Yet these images tell only a part of the story. Where we purchase food and where it comes from (in particular, its geographic origin) does not always reveal how local food gets to the point of sale or why it is sold touting some characteristics and not others. The promotional flyer we might read at the supermarket meat counter about the nearby farmer of grass-fed beef likely does not describe the importance of interdependent business relationships between the farm, slaughterhouse, and retailer. Neither does the bin full of the season's first apples at the farmers market tell you about the grower's significant investment in transportation and marketing activities that allow him to sell in multiple markets each week.

The stories behind the images describe the people, processes, and relationships—that is, all the segments of the supply chain—that put local foods into consumers' hands. The supply chains for local foods, like those of more mainstream products that account for the vast majority of food consumed in the United States, remain largely hidden from consumer view. Yet it is within these supply chains that the food characteristics and the information consumers value are determined. As local foods become a more important part of the U.S. food system, our understanding of the inner workings of food supply chains deserves more attention.

A Growing Trend in Food and Agriculture

U.S. consumer interest in local foods has increased sharply in recent years. Although sales of locally grown food still account for only a small share of total domestic food sales, this is believed to be one of the fastest growing segments of the U.S. food system. Interest in local foods stems from a variety of potential and perceived benefits, including economic, environmental, health, food safety, and rural development benefits. Some believe that local food supply chains provide several advantages over the mainstream supply chains that provide products to supermarkets. These might include preserving local landscapes and family farms, strengthening of local and regional economies, and providing fresher, higher quality food products. Certain consumer segments are actively seeking local foods in a variety of outlets, and there is evidence from anecdotal observations and from controlled economic experiments that some consumers are willing to pay higher prices for local foods.

These trends are prompting changes across a spectrum of food supply chains. Farmers' increased utilization of direct marketing channels such as farmers markets and a variety of community supported agriculture business models is providing an important market mechanism linking farmers and consumers.[1] Some argue that direct market channels give farmers more control over distribution and allow them to capture a higher share of retail value in comparison to selling through mainstream intermediaries. At the same time, these channels offer an alternative outlet for consumers to seek local fresh products directly from the source. But direct marketing channels are not the only channels through which locally grown foods are made available to consumers. A number of mainstream supermarkets, which are remarkably resilient and quick to adapt, see these trends as an opportunity to satisfy customer demand for local foods and to increase customer loyalty. However, it is not clear that this is an effective channel for meeting the rapidly growing demand for local food products; and there is uncertainty about the long run prospects for a significant "re-localization" of supermarket offerings.

Interest also extends to federal, state, and municipal policymakers, who seek to marshal significant resources to support local food systems. Local foods are increasingly being incorporated into programs to reduce food insecurity, support small farmers and rural economies, improve healthy eating habits, and foster closer connections between farmers and consumers. Local governments, for example, are implementing an array of training programs for vendors and farmers market managers to improve skills in running local food supply chains. Municipalities are also making capital investments in infrastructure to facilitate the development of supply chains for local foods. Today many states and cities have food policy councils centered on promoting local foods. In addition, there is strong interest in increasing the share of local foods, in particular fresh fruits and vegetables, in the federal Supplemental Nutrition Assistance Program (SNAP). The U.S. Department of Agriculture (USDA) administers several grant and loan-guarantee programs that can potentially support local food supply chains. In the regulatory arena, an emerging issue is the differential treatment regarding food safety and product traceability that the federal government uses with direct market supply chains relative to their mainstream counterparts.

Despite the growing importance of local food supply chains to consumers, food supply chain members, and policymakers, relatively little is actually known about them. Nor is the performance of local food supply chains well understood in terms of economic, human health, environmental, and social effects. To understand the local food phenomenon better, this book offers a rigorous comparison of local and mainstream supply chains in multiple social, economic, and environmental dimensions, using the case study method. The fifteen case studies and the systematic comparison of case study findings are intended to shed light on the factors that will influence the structure, size, and performance of local supply chains in coming years.

What Can We Learn About Local Food Supply Chains?

A consequence of the rapid growth in local foods is that our understanding of these marketing arrangements has struggled to keep pace with interest in them. In some ways, this makes for a more interesting environment to study. Producers large and small are trying new strategies to enter different markets or create new market niches where none existed before. Consumers are seeking a broader array of products and product characteristics and are increasingly open to obtaining food through different market outlets. This results in an abundance of new and creative approaches to food supply chains, some more successful than others and many not yet well understood.

Much has been written about the different examples of local food marketing, the demands and motivations of consumers, and the potential costs and drawbacks of an expanding local foods sector. There have also been numerous attempts to probe the potential of local foods, or other alternative food marketing arrangements, to contribute to a more just and sustainable food system. Yet our current understanding of local foods often either ignores how food moves from producers to consumers or makes assumptions about this process that may not be realistic.

In this volume we attempt to "catch up" with the growth in local foods by better understanding how local foods move from producers to consumers. We focus on the relationships and arrangements that make up food supply chains, observe how they are organized when food moves through local supply chains compared to mainstream supply chains, and ask why these arrangements may differ between supply chain types.

At the most basic level, these observations help illustrate how local foods are being introduced or reintroduced into the broader food system. For example, do we tend to observe local foods moving through entirely new pathways from producers to consumers, or do they rely on existing infrastructure, knowledge, and relationships (which may have been developed for more mainstream supply chains) to provide consumers with the food and product characteristics they demand? The answer to this type of question has implications for how we think local foods will develop in the future.

Understanding the "how" and "why" of local food supply chains also provides a basis for answering some of the thornier questions about local foods, such as whether barriers to growth in local foods exist (and why they might exist), and how barriers could be removed. In all sorts of markets there are barriers and costs to entry, and there are structural market forces that determine who participates and what supply chains look like. These features depend on relationships between supply chain partners upstream and downstream from any one entity. Yet we know little of these relationships as they pertain to local food supply chains.

Other topics are increasingly making their way into the public discourse on local foods, with a particular focus on the role of local foods in public policies and programs. Questions about measures of supply chain performance, such as the prices and availability of local foods, the potential of local foods to provide public benefits (like improved environmental quality or better nutritional outcomes), and the prospects for growth in local foods are difficult to answer for a segment of the food sector that is growing so rapidly.

Of particular difficulty is answering the question of why we might think that local foods, and the supply chains that deliver them to consumers, differ from other marketing arrangements in how they perform. That is, what aspects of local food supply chains suggest that they may provide certain public benefits, and what does this mean for public policies and programs designed to support local foods? Our goal is to use observations from supply chain case studies to begin to describe better the distinguishing characteristics of local food supply chains and to ground the resulting discussion more concretely in how food supply chains operate.

Using Case Studies to Understand Supply Chains

One of the reasons why little research has focused on supply chains for local foods is that information on supply chains is sparse and difficult to gather. Most research on food and agriculture systems uses data sources that describe individual segments of supply chains. For example, the U.S. Department of Agriculture gathers comprehensive information about activities on the farm using the Census of Agriculture and the Agricultural Resource Management Survey. Data from these sources also provide some information about how food (and non-food agricultural products) is marketed. Yet these and other sources focused on individual entities in the supply chain have difficulty describing how an entire supply chain operates.

In this volume we use a series of case studies to understand better how local food supply chains operate.[2] We examine supply chains for five food items, one each in five different locations. We study supply chains for apples in Syracuse, New York; blueberries in Portland, Oregon; spring mix in Sacramento, California, beef in the Twin Cities, Minnesota, and milk in Washington DC. Each of these product-place combinations can yield unique insights into how food—whether local or not—reaches consumers. These product-place combinations were selected to allow for rich comparative analysis across products and geographic locations. The five products represent both plant- and livestock-based foods. Blueberries and spring mix are sold fresh and are highly perishable. Apples are also sold fresh but can be stored effectively and sold year-round. Beef and milk require processing and are available year-round. The five places are geographically dispersed and differ considerably in population and per capita income. Additional study locations in the Southeast and Southwest would have added still more diversity to the food supply chains investigated, but resource constraints limited the number of locations that could be studied.

The case study method is used for two complementary reasons. First, we use the entire supply chain (at least, up to the point of sale) as the unit of observation, and we gather information about how products move from producers to consumers in different types of supply chains. A case study format, where we gather detailed information across many dimensions, allows us to examine how all of the individual segments fit together to form a supply chain.

The second reason for using case studies is that the questions we ask about supply chains are not easily answered using more traditional representative data sets. We are interested in the how and why of supply chains and in probing the great variety of ways in which local foods reach consumers. However, this type of information is not readily gathered through typical survey or census forms. We seek information on relationships and arrangements between supply chain partners and details about how many moving parts work together in a supply chain. A detailed case study can provide a more complete picture of the entire supply chain, although by design a case study cannot gather this information for a representative sample of supply chains.

Case studies of supply chains are appropriate for studying local foods because of the significant information component that is necessary to distinguish local from nonlocal. In developing this series of studies, we recognize that *local* is not adequately defined by geographic boundaries (e.g., within a state or region) or by the type of market where local foods might be found (e.g., farmers markets or community supported agriculture programs). Instead, consumers may seek a connection to where, how, and by whom their food was produced, information that is difficult to obtain through mainstream supply chains. To get this connection requires information to be conveyed through the supply chain.

We hypothesize that how information is conveyed is an important factor in determining what a supply chain looks like and how it operates. Under this hypothesis a farmers market, for example, may not simply be a way for consumers to access farm-fresh products from many producers. It is also a vehicle for conveying detailed information directly from those who produced the food. But other types of supply chains—both local and nonlocal—may convey different types of information and require a different supply chain structure. Case studies can help us learn more about how information about a product's origin shapes how it is delivered to consumers and why we observe differences in structure, size, and performance across supply chains.

Key Themes about Local Food Supply Chains

To understand the local food phenomenon better, this volume offers a rigorous comparison of local and mainstream supply chains using the case study method. The book centers on two broad questions. First, what factors influence the structure and size of local food supply chains? Here "structure" refers to the configuration of processes, participants, and product flows as a product moves from primary production to consumers. "Size" refers to aggregate sales volume as a percentage of total food sales for a product category. Second, how do local food supply chains compare to mainstream supply chains for key dimensions of economic, environmental, and social performance? By exploring these questions we can increase our understanding of the ways in which food products can move from farmer to consumer, and we can reveal more nuanced supply chain relationships than are commonly recognized in the public discourse on local foods.

Several themes emerge as the case studies address these broad questions. One theme centers on identifying the flow of local foods and the corresponding volumes in each channel through which they reach consumers. The interest here is in determining the amount of local foods in direct market chains and/or intermediated chains (both mainstream and other types of intermediated supply chains). The common perception is that local, direct market supply chains tend to be smaller. It is possible that limited access to processing and distribution, effects of public regulations and commercial business policies, and a lack of year-round supply may hinder growth prospects for local products. The cases presented investigate whether local supply chains have the ability to connect with the mainstream supermarket supply chain, offering opportunities for the expansion of the supply of local foods.

A related issue is that distribution services are handled internally by direct market producers and likely require considerable resources. Direct market producers may spend considerable time packing, driving, and selling at farmers markets. Moreover, distribution costs per unit of product may be higher in local intermediated supply chains relative to those in mainstream chains. As demand for local food products grows in an area, questions arise regarding processing and distribution capacity. The reason is that intermediated chains carrying local products may need to grow product volumes to the point where they are large enough to gain size economies by distributing through mainstream distribution centers.

A second theme highlighted in this book is the role of information flows for local foods. Although local foods may be expanded by connecting local growers to the supermarket supply chain, the presence of intermediaries may complicate establishing and maintaining strong connections for consumers about where, how, and by whom their food was produced. This is particularly relevant for consumers who value the connection with the producer and the perceived impacts of local foods on local economies. In direct market chains this information is easily conveyed because it is part of the shopping experience, whereas facilitating such information flow in intermediated chains may require additional effort, costs, and coordination among supply chain members.

A third theme in this book is the identification of business models employed by successful local food supply chains, highlighting their similarities and differences with respect to mainstream chains. In particular, the cases studies allow us to examine whether supply chains that carry local products tend to share common characteristics and to identify the portfolio of marketing channels of producers selling local foods. Other elements that affect the performance of local supply chains are considered in the cases, including the quality of relationships among supply chains members (e.g., trust, information and decision sharing), the price-making decision process, and the engagement of collective organizations, among others.

The fourth theme centers on the study of interrelated economic, environmental, and social performance elements of local supply chains using the mainstream channel as a benchmark for comparison. One expects, for example, that producers in local food supply chains tend to receive a larger share of the retail price. But this comparison must take into account that in direct market chains the producer has to assume greater responsibility for the supply chain functions, which are often costly. This book sheds light on whether producers are financially better off in local supply chains, as this depends not only on the share of retail value but also on the volume of sales, the size of the price premium they receive, and how cost-effectively they can perform additional supply chain functions. Likewise, one expects that longer distance traveled by products implies a higher per pound fuel consumption, suggesting that local supply chains may perform better than mainstream chains in terms of fuel use efficiency. Nevertheless, it is possible that reduced per-unit costs achieved through the product aggregation practiced by mainstream supply chains outweigh cost differences due to proximity to the consumer.

Finally, a fifth theme throughout the book is the role of public policies in shaping the structure of food supply chains. Food safety assurance, commercial operating standards, and mandatory certification systems may affect smaller food businesses more than larger ones. At the same time federal, state, and local governments are designing policies and implementing programs to facilitate the development of local supply chains. The case studies reflect the perception of growers, processors, distributors, and retailers of local and mainstream supply chains alike, regarding the impact of these and several other policy issues that are shaping the structure, size, and performance of local supply chains.

Exploring these themes provides a detailed account of how food in local and mainstream supply chains reaches consumers. In the next chapter we present the theoretical underpinnings of our research hypotheses and the research design that guided the study. The following five chapters present in detail the supply chain case studies conducted in each location. These chapters comprise the empirical core of the book; many of our insights and conclusions are derived from observations and cross-case comparisons using the supply chain case studies. Chapter 8 analyzes the importance of supply chain characteristics in determining the price and availability of the products studied in the previous chapters. Of particular concern is gaining a better understanding of how products in local supply chains are positioned in the marketplace.

Chapters 9 through 11 summarize what we have learned in conducting the case studies and present our insights on the themes mentioned. Each of these essays addresses a broad topic on the current state and future of local foods, with insights drawn from the case studies and elsewhere. The first, titled "What Does Local Deliver?" asks what changes we might expect from a relatively small re-localization of the food system. The second, "Can Local Food Markets Expand?" takes what we have learned about the barriers to and opportunities for expansion of local foods and projects how local food markets might develop from here. The third essay, "What Role Do Public Policies and Programs Play in the Growth of Local Foods?" examines how public interventions related to local foods may or may not influence their development.

Taken together, the case studies and other analyses begin to indicate the way forward for local food supply chains and their place in the U.S. food sector. In particular, they suggest that food supply chains are increasingly adaptable to new consumer demands and that there is no one-size-fits-all supply chain type for local foods. Local foods are clearly commanding increasing attention in the food marketplace, and there is no sign that activity and interest in local foods is dwindling. However, the expansion path is uncertain and highly dependent on consumer tastes and trends, food and agriculture policy, and myriad economic forces at local, national, and global scales.

Notes

1. Community supported agriculture generally refers to arrangements where consumers and producers share the risks and benefits of food production (Holcomb et al. 2013). These can take many forms. A common example is a subscription service, where consumers pay a flat fee at the beginning of the season that entitles them to a share of the available product.

2. The research for the case studies was funded under a series of cooperative agreements between USDA's Economic Research Service and four participating universities: Cornell University, Oregon State University, the University of California–Davis, and the University of Minnesota. Results from that research were originally published in King et al. (2010).

References

Holcomb, Rodney B., Marco A. Palma, and Margarita M. Velandia. 2013. "Food Safety Policies and Implications for Local Food Systems." *Choices* 28(4).

King, Robert P., Michael S. Hand, Gigi DiGiacomo, Kate Clancy, Miguel I. Gómez, Shermain D. Hardesty, Larry Lev, and Edward W. McLaughlin. 2010. *Comparing the Structure, Size, and Performance of Local and Mainstream Food Supply Chains*. ERR-99. Washington DC: U.S. Department of Agriculture, Economic Research Service.

2   Research Design for Local Food Case Studies

Robert P. King, Michael S. Hand, and Gigi DiGiacomo

This chapter begins with a review of conceptual foundations for studying local food supply chains and presents a series of specific research questions that form the basis of our case study analysis and data collection. The discussion of conceptual foundations centers around four issues: (1) defining *local*; (2) supply chain structure; (3) supply chain size; and (4) supply chain performance. The chapter concludes with an explanation of the unit of analysis for the case studies and a description of the overall research design for the study, which can be classified as a multiple case design.

Conceptual Foundations—Defining *Local*

Despite its growing use in academic and civic discourse, there is no consensus on a precise definition of *local*. Clearly the term refers to a place that is circumscribed by geographic boundaries. However the relevant boundaries for what consumers perceive to be local may vary considerably across locations and among products. Results reported by Durham et al. (2008) show that there is no clear consensus among consumers in a particular locale as to the definition of *local*; that the average radius of the area designated to be local varies considerably across regions; and that the radius of the local area is larger for processed products than for fresh fruits and vegetables. Findings from this study—and from Ostrom (2007) —also show that definitions of *local* based on state boundaries fail to capture many consumers' beliefs.

Many consumers also link production practices, cultural values, and the geographic scope of distribution to their concept of local. For example, "sustainable" production practices and "family farms" are often associated with local products, though these added attributes are usually not clearly defined.[1] Similarly, products that are produced locally but distributed nationally may not be perceived as local products.

The concept of a "Short Food Supply Chain" (SFSC), which Marsden et al. (2000) characterize by an emphasis on a connection between the food producer and food consumer, is especially useful for defining local in the context of this study. They note:

> With a SFSC it is not the number of times a product is handled or the distance over which it is ultimately transported which is necessarily critical, but the fact that the product reaches the consumer embedded with information, for example printed on packaging or communicated personally at the point of retail. It is this which enables the consumer to confidently make connections and associations with the place/space of production *and, potentially, the values of the people involved and the production methods employed.* The successful translation of this information allows products to be differentiated from more anonymous commodities and potentially to command a premium price if the encoded or embedded information provided to consumers is considered valuable. (Marsden et al. 2000: 425, emphasis in original)

They go on to identify three main types of SFSCs: (1) face-to-face chains with direct purchases from farmers, (2) spatial proximity chains that make consumers aware of local origin at the point of purchase, and (3) spatially extended chains that convey the value and meaning of a place of production to consumers outside the region where the product is produced. The first two of these are especially relevant in the U.S. context, while the third is more relevant for Europe.

For this study we were flexible in developing a working definition of *local*, being open to refinements based on findings as the case studies progressed and to the possibility of having a different definition of *local* for each case study location and product. As a starting point, we defined *local products* as those originating within the study area state or within the study area state and the surrounding states. We remained flexible as to whether "origination" referred to farm production or processing. In the case of meat products, this was an important distinction. Also, we were flexible as to whether to exclude products produced locally but distributed nationally—for example, spring mix from large growers in California or blueberries from Oregon during some months. Finally, we were open to allowing the definition of *local* to vary across product categories in our case studies. *Nonlocal* products were defined simply as those that are not locally produced.[2]

In the case studies presented here, we define a local food product as one raised or produced and processed in the locality or region where the product is marketed. For each of the five study locations, we establish a geographic area that defines the locality or region where local food products originate. These are:

Syracuse, New York: New York State.
Portland, Oregon: Oregon and Washington.
Sacramento, California: Sacramento Metropolitan Statistical Area (MSA), made up of El Dorado, Placer, Sacramento, and Yolo Counties.
Twin Cities, Minnesota: Minnesota and Wisconsin.
Washington DC area: Washington-Baltimore-Northern Virginia Combined Statistical Area, made up of the Baltimore-Towson, Maryland MSA; the MSAs of Culpeper, Virginia, and Lexington Park, Maryland; the Washington-Arlington-Alexandria MSA, spanning parts of DC, Maryland, Virginia, and West Virginia; and the Winchester MSA, in parts of Virginia and West Virginia; plus the counties immediately adjacent to (i.e., sharing a border with) the combined statistical area.

We also distinguish between local food products and local food supply chains. A *local food supply chain* is defined as the set of trading partner relationships and transactions that delivers a local food product and conveys information to consumers about where, by whom, and how the product was produced. This definition implies that the supply chain conveys information about the product enabling consumers to recognize it as a local food product, even when the producer and consumer are separated by intermediary segments in the supply chain.

Conceptual Foundations—Supply Chain Structure

Economic theories of vertical integration and optimal ownership help identify factors that affect the structure of supply chains. Williamson's (1975, 1986) work on transaction cost economics is useful for understanding supply chain structure—especially for understanding whether it is more efficient for transactions along a supply chain to be governed by market relationships between distinct firms or to be internalized within a single firm. Vertical integration can be a response to high transaction costs, especially those associated with hold-up problems linked to asset specificity. Direct-to-consumer supply chains are an example of vertical integration, since the producer who sells directly to consumers internalizes distribution and retail activities that are performed by independent firms in mainstream retail channels for most food products. Direct marketing and short intermediated local food supply chains help producers create and capture price premiums associated with unique attributes of their products. This comes at a cost, however, since the optimal scale for some activities may be much larger than the scale that can be achieved by a single producer. Mainstream supermarket and food service supply chains may sacrifice value created by a direct producer-consumer link, but this loss can be offset by economies of size and by reliance on low prices that emerge when products change ownership in highly competitive markets. As local food supply chains scale up, it will be important to determine how significantly ownership changes in intermediated chains affect the information provided to consumers on where, by whom, and how products are produced.

Williamson's work also suggests that stable, long-term relationships with trading partners or service providers are another structural response to asset specificity. In such relationships, trading partners are willing to forgo short-run price opportunities offered by firms outside their trading network and may base transaction prices on shared perceptions of long-term production costs rather than on competitive market prices. In effect, this decouples transaction prices from short-term fluctuations in supply and demand. Firms establish these "relational contracts" because they derive significant benefits—either enhanced product differentiation or significant logistics or transaction cost savings—from their long-term association. This raises the question of whether durable relationships between supply chain partners are more common and more important in local food supply chains and, if so, whether these relationships facilitate the decoupling of local food prices from commodity prices.

Building on the work of Williamson and on property rights theories developed by Hart (1995), Hansmann (1996) asserts that economic activities will tend toward organizational structures that minimize the combination of contracting costs and ownership costs and that ownership will be concentrated in segments of a supply chain that make unique contributions to value creation. Producer marketing cooperatives and consumer cooperatives are two of the organizational forms he considers.

Hansmann (1996) argues that producer marketing cooperatives are most likely to form when farmers acting independently are subject to monopsony purchasing power from a single buyer—such as a processor or wholesale distributor—and when the costs of collective decision making are reduced by physical proximity, shared values, and common interests. Producers who sell direct to consumers have multiple buyers and so do not face monopsony power, but the formation of marketing cooperatives may be worthwhile for local food producers who aggregate their product and sell to market intermediaries such as supermarkets or food service distributors.

Hansmann (1996: 161–62) also observes that retail cooperatives are relatively rare, noting: "The costs of customer ownership for many retail goods and services are high, because the customers of any given retail firm are commonly too numerous, transitory, and dispersed to organize easily and effectively. The costs of market contracting are commonly low: retail markets for most ordinary items are sufficiently competitive to keep prices close to cost, and the goods and services themselves are sufficiently simple or standardized, or are purchased so repetitively, that asymmetric information about quality is not a serious problem." In local food supply chains, however, customers often share common values and seek a stable rather than transitory relationship with producers. The potential for problems stemming from asymmetric information can be significant in local food supply chains, where many products are credence goods with attributes that cannot necessarily be independently verified. This suggests that consumer cooperatives may be more prevalent in local food supply chains than in mainstream food supply chains.

Finally, while businesses or even entire supply chains may compete with one another, they often benefit from having shared access to physical, institutional, and informational infrastructure that is made possible by having a critical mass of users. For example, trade associations provide access to information and potentially valuable lobbying services for their members, at either a state or national level. Similarly, farmers markets and public and private sector programs that promote local foods can be viewed as infrastructure that helps connect competing local food product suppliers with potential customers. Linkages to the knowledge and service infrastructure created by a strong industry that distributes nationally or internationally may be of critical importance for some local food supply chains. Alternatively, the development of infrastructure for one or more strong, well-established local food supply chains may be more important for the development of emerging local food supply chains.

Conceptual Foundations—Supply Chain Size

The size of local food supply chains, as measured by sales volume relative to total demand, is likely to be limited by a focus on selling within a circumscribed region and the desire to foster strong linkages between producers and consumers. These factors may limit both supply and demand. If the supply of local food products exceeds the demand for them at observed transaction prices, local food product producers will sell excess production into commodity markets, and observed prices in local food product markets will not be significantly above commodity market prices. On the other hand, if demand for local food products exceeds supply, local food producers are likely to sell all their production in local food markets, and observed prices for local products will be well above commodity prices.

Constraints associated with processing and distribution activities may also affect sales volume in local food supply chains. These activities often have relatively high fixed costs, and costs per unit of food processed or transported are often quite sensitive to scale of operations and capacity utilization. With relatively small volumes of product flowing from farms, lack of access to and costs associated with processing and distribution services may limit the size of local food supply chains. Similarly, high processing and distribution costs may raise product costs to the point where it is difficult for local food producers to compete on price with products in mainstream supply chains.

Public policies and commercial regulations can also impose significant costs for compliance that are often relatively insensitive to size of operation. In effect, these are fixed costs that disproportionately increase unit costs for low-volume enterprises. These fixed costs for compliance with regulatory and operating standards (public or private) may limit the ability of local food products to compete with products in mainstream supply chains.

The focus on production location also limits supply chain scope, as measured by seasonal availability for fresh products. Of course, this also may limit potential sales volume of local food products relative to sales in mainstream channels, since year-round product availability enabled by sourcing from multiple locations is a key element of the business strategy for these businesses.

Finally, as local food supply chains grow in response to increasing demand, their mode of growth will shape the overall structure of the local food system. On the one hand, growth may occur through replication—that is, through the entry of new firms. This fosters innovation and competition and may make it easier for producers and consumers to establish and maintain close, personal relationships. At the same time it may be difficult for small firms to take advantage of potential economies of size in some supply chain activities because of large capital investment requirements. Alternatively, growth may occur through internal expansion of existing firms. Often internal growth is driven by size economies, but in some cases such growth requires fundamental changes in business processes that can be costly and can fundamentally change relationships with customers. There is a need to understand better the factors that determine whether growth is more likely to occur through replication or through internal expansion.

Conceptual Foundations—Supply Chain Performance

The equitable distribution of revenues and costs across the supply chain is a key concern in supply chain design and an important aspect of supply chain performance. This is a central theme in the literature on "lean" supply chains (Womack et al. 1991; Womack and Jones 1996; King and Venturini 2005). It also plays a key role in more recent writing on values-based supply chains. As Stevenson and Pirog (2008: 131) note: "Strategic partners are rewarded based on agreed-on formulas for adequate margins above production costs and adequate returns on investment. This differs markedly from low-cost bidding mechanisms," which govern most transactions in the predominant supermarket and food service supply chains. Local food advocates have placed particular emphasis on returns to farmers that cover their production costs and on the total share of consumer expenditures that goes back to farmers, but it is important to understand whether farmers who sell directly to consumers are actually better off after netting out the extra marketing costs they assume.

It is commonly perceived that local food products are more expensive, and a higher price may be necessary if producers are to capture higher per unit revenues and a greater share of the consumer dollar. Often, however, price premiums may be due to unique product attributes that are quite distinct from being local. This suggests that the attribute "local" may not be enough to command a significant price premium. Rather, differentiation by other quality attributes that require extra efforts or unique capabilities may be necessary to receive and sustain price premiums for local food products.

The desire to keep more money circulating in the local economy is often cited as a reason for consumers to "buy local." Local food supply chains are often more labor intensive and less capital intensive than mainstream supply chains, and labor utilization in local food supply chains often is more concentrated in upstream farm production and processing enterprises. Since local food supply chains, by definition, operate in more circumscribed areas, the labor they use is more likely to be based in the local economy. This could imply important local economic impacts. On the other hand, the distribution and retail components of the mainstream food supply chains are also relatively labor intensive, and effective wage rates for local employees in these businesses may be higher than those in businesses that supply local food products. Therefore, it is not clear that the concentration of costs for employee and proprietor labor inputs in the farm and processor segments of local food supply chains actually leads to a larger overall contribution of wage and business proprietor income in local economies.

Energy use for transportation of food is also a matter of growing public concern. Local food advocates have used the concept of food miles as the basis for arguments that local food systems are more energy efficient. Local food products almost certainly travel fewer miles than comparable products sourced in from other regions or other countries. Mode of transportation also matters, however, and the predominant supermarket and food service supply chains are often able to realize significant energy efficiency gains by using more efficient modes of transport that take advantage of scale economies.

Finally, advocates of local foods have asserted that increased availability and consumption of local foods helps create a stronger sense of community and civic engagement. For example, Lyson (2007: 19) states: "The new paradigm, labeled civic agriculture, is associated with a relocalizing of production. From the civic perspective, agriculture and food endeavors are seen as engines of local economic development and are integrally related to the social and cultural fabric of the community. Fundamentally, civic agriculture represents a broad-based movement to democratize the agriculture and food system."

There is a need to investigate the extent to which local food supply chains do more than their mainstream counterparts to foster the creation of social capital and civic engagement in local communities.

Research Questions

The inquiry process for each case study was designed to answer the two broad questions that are the focus for the overall study:

What factors influence the structure and size of local food supply chains?

How do existing local food supply chains compare with mainstream supply chains for key dimensions of economic, environmental, and social performance?

The following specific research questions, which emerged from the preceding discussion of conceptual foundations, served as the guide for case study and data collection analysis.

SUPPLY CHAIN STRUCTURE

Do direct and intermediated food supply chains provide the consumer with detailed information about where, by whom, and how the product was produced?

Are durable relationships between supply chain partners—characterized by a high degree of trust, information sharing, and decision sharing over time—important in food supply chains where trading partners exhibit strong mutual interdependence or one partner depends on another in a unique way?

Are prices in direct and intermediated food supply chains decoupled from prices determined in commodity markets?

What is the role of collective organizations (such as producer and consumer cooperatives and farmers markets) in direct and intermediated food supply chains?

Does the presence of a strong industry that distributes nationally or internationally help create an infrastructure of knowledge and services that facilitates the development of direct and intermediated food supply chains?

Does the presence of local food supply chains for other products and broader local food initiatives help create an infrastructure of knowledge and services that facilitates the development of other successful direct and/or intermediated food supply chains?

SUPPLY CHAIN SIZE

What is the portion of total demand in a general product category represented by products sold in direct and intermediated food supply chains?

Do problems with access to and costs associated with processing and distribution services limit the size of direct and intermediated food supply chains and raise product costs to the point where it is difficult to compete with products in mainstream food supply chains?

Do fixed costs for compliance with regulatory and operating standards (public or private) limit the ability of low-volume local food products to enter mainstream supply chains?

Does lack of year-round availability limit market opportunities for local food products?

Do direct and intermediated food supply chains respond to growth opportunities through replication of firms or through internal firm expansion?

SUPPLY CHAIN PERFORMANCE

After subtracting marketing costs, do producers receive higher per unit revenue and retain a greater share of the price paid by the final consumer in direct and intermediated food supply chains?

Is differentiation by quality attributes other than "local" that require extra efforts or unique capabilities necessary to receive and sustain price premiums for local food products?

Does concentration of costs for employee and proprietor labor inputs in farm and processor segments of direct and intermediated food supply chains result in a larger contribution of wage and business proprietor income to local economies?

Does a typical unit of product in direct and intermediated food supply chains travel fewer miles and use less fuel for transportation per unit of product sold?

Do direct and intermediated food supply chains foster the creation of social capital and civic engagement in the consumption area?

Unit of Analysis

The basic unit of analysis for this study was a single supply chain—for example, apples in the metropolitan area of Syracuse, New York, that are marketed through a direct marketing supply chain. A supply chain is composed of a series of technologically distinct processes that extend from the provision of inputs for primary agricultural production through production, processing, distribution, and sale to an end user. The physical and institutional infrastructure that supports these processes is also considered to be part of the supply chain. The description of a supply chain identifies groups of participants and their scope of activities, the magnitude and timing of product flows, the incidence of costs and returns within the chain, patterns of communication and information exchange, the allocation of property rights and decision authority, and a set of performance metrics such as price paid by consumers, energy usage, and total employment. In total, the study observed fifteen different supply chains: three each for five product-place combinations.

Study Design

This study is based on a multiple case design, with the cases clustered along two dimensions (Yin 1994). First, cases were clustered by product-place combination. Three case studies were conducted in each of the five study locations. As already noted, they focused on mainstream, direct market, and intermediated supply chains for a single product category. The cases for each location shared a common physical, economic, and institutional context that shapes the overall food system in the location. This clustering facilitated direct comparisons among the three chain types. Second, the cases were clustered across locations by chain type. This clustering allowed for comparisons of similar chain types across products and locational contexts in order to determine whether it is possible to make general conclusions about factors associated with the structure, size, and performance of local food product supply chains.

The product-place combinations for this study were selected to permit rich comparisons across products, geographic locations, and supply chain structures. The five study locations were identified first. They differ widely in population and per capita personal income. All are metropolitan statistical areas (MSA). As such, they are centers of population and commercial activity. This made it easier to study direct marketing supply chains, since the greater concentration of consumers in urban places allows farms engaged in direct marketing to reach a critical mass of potential customers. The five products were identified after the study locations. They include foods based on crop and livestock production and both processed and unprocessed products. In order to study intermediated supply chains, which generally require significant volumes of product, there needed to be a certain critical mass of local production for each of the products selected. A significant level of local production is also needed to support processing infrastructure for products like beef and milk.

This study design imposes at least two limitations that should also be noted. First, the focus on MSA locations limits the opportunity to study local food supply chains in rural areas. The focal firm for a pilot case study used to develop interview and reporting procedures for the case studies included in this book was a small cooperative that aggregates local production over a diverse set of product categories and sells in both rural and urban areas.[3] The fact that this cooperative has great difficulty in selling to rural customers points to the need to understand better how local food supply chains operate in rural communities. The second limitation stems from the requirement that there be a critical mass of local production for the products selected for study. This precludes study of emerging local food supply chains that are established in response to unmet consumer demand for a particular category of local food products. These supply chains face a number of challenges that are worthy of study, especially for products that require processing.

Field Procedures

We used three techniques to collect data for this study: (1) interviews with supply chain participants, (2) direct observation of product availability and prices in various market settings, and (3) collection of contextual data on current and past food system characteristics. General procedures for each follow.

The case study interviews, direct observation of prices and availability, and collection of contextual data were conducted during 2009. Observations and conclusions reflect the situation at that time and have not been updated to account for subsequent events and developments.

Interviews with Supply Chain Participants

Personal interviews were the primary source of information on supply chain structure, size, scope, and performance. These were time-consuming for research team members to set up, conduct, and analyze. They also required a significant time commitment from supply chain participants. The following general procedures were used for these interviews.

Learn as much as possible about the supply chain to be studied by visiting web sites, reading articles in the trade press, and/or talking with industry experts.

Identify the right people to interview. It was usually best to start by identifying the person or organization playing a leadership or coordinating role in the supply chain. The first contact could be a farmer, the manager of a cooperative or branded food company, or the manager of a food retailing firm. Having a key contact of this kind agree to participate in the study meant the individual could be very helpful in identifying and making initial contact with other supply chain participants.

Make initial contact in person or by phone, if possible, though initial contact could also be made by email or letter. Information shared at this time included a brief overview of the project, an explanation of why the supply chain in which this person or firm participated would be a good subject for a case study, and a preliminary request for participation. If the contact was made in person or by phone, project personnel followed up with a brief email or letter that included a copy of the project summary. In some cases, the initial contact was not the person who should be interviewed. Project personnel were prepared to speak with several people, especially in a large organization.

Make a more formal request for participation. Normally, this was done in a short phone call that included a description of the range of questions to be asked in the interview and an initial discussion of confidentiality procedures.

Schedule the interview. Normally this was at the workplace of the interview subject. The time required for the interview depended on the type of organization and the questions to be asked but normally ranged from one to two hours.

Prior to the interview, send a confirmation note that may include a listing of the questions to be asked and/or reminders about information the subject might want to have available during the interview.

Have two research team members present, one to take the lead on asking questions and the other to take notes. Interviews could be tape recorded, but this could intimidate some subjects. Notes were critical, even if a session was recorded.

Begin the interview with any necessary introductions, a brief overview of the project, and a reminder about confidentiality procedures.

In general, follow the line of questioning specified in the interview question guide for the type of firm being interviewed.[4] Project interviewers were prepared to deviate if unexpected issues arose.

When possible, collect sample documents and/or quantitative data. If not available during the interview, ask if these can be provided later.

Conclude the interview with a short discussion of the timeline for feedback and follow-up questions. If this was the first interview for a supply chain, project interviewers asked who should be interviewed in other supply chain segments and requested the help of those being interviewed in contacting those people and encouraging their participation.

If possible, tour the subject's workplace after the interview.

Send an email or note of thanks.

Prepare a written summary of the interview as soon as possible. This summary was reviewed by each of the team members who participated in the interview to ensure accuracy and to identify any need for follow-up questions.

Ask any necessary follow-up questions by email or by phone.

Prepare a written draft of the case study after all participants in a supply chain have been interviewed. This draft followed the general structure used in the pilot case study as closely as possible.

Send relevant sections from the draft case study to each interview subject for review. The draft text was accompanied by a release form. Those being interviewed were asked to return the release form along with comments and suggestions.

Incorporate comments and suggestions from study participants. After editing the case study, project interviewers sent a copy of the entire draft case study to each participant for review.

Direct Observation of Product Availability and Prices

Data on product availability and prices were collected weekly throughout 2009 for each of the five products in each of the five study locations. In general, market locations included two supermarkets, two natural foods stores, and two farmers markets. Although these market locations were typical of a wide range of retail outlets for the products studied, they may not have been fully representative of all potential outlets for local and nonlocal food product supply chains. A "recorders' guide" was developed to outline data collection techniques and to help facilitate consistency across all study locations.[5]

Contextual Data on Food System Characteristics

Geographic, demographic, economic, and agricultural sector data for the period 2007–2009 were collected from secondary sources for each product category and metropolitan location to provide a context for the case studies and cross-case comparisons. Data from each location concerning metropolitan population, per capita income, commodity production, and direct marketing, for example, were compared to national averages. This provided insight into the size of the local market, consumer purchasing power, the economic importance of the products studied, and activities related to marketing of local foods. Sources for these data included the U.S. Census of Agriculture, USDA Economic Research Service, USDA Agricultural Marketing News Service, and USDA National Agricultural Statistics Service. Other third-party sources of information such as websites and trade-related articles were used to document trends in distribution and retailing.

Analysis Procedures

The overall project team met face to face three times: in December 2008, when an initial draft of the research design had been prepared; in June 2009, when interviews with case study participants and secondary data collection were under way; and in December 2009, when drafts of all the chapters for all five of the product-place combinations had been completed. Face-to-face meetings were supplemented with monthly project team conference calls.

The project team developed common procedures for describing supply chain structure and guidelines for presenting background information on study location and on general product characteristics. The team also developed standard analysis methods and table formats for some of the quantitative information presented in each case study—for example, product availability charts, revenue allocation analysis, and food mile and transportation fuel use analysis. Although common procedures were used, the team allowed some variation across locations and individual cases in order to highlight unique findings or interpretations of the data. Analysis of the data collected in each study location was done by the team members who conducted the case study participant interviews and collected secondary data.

Notes

1. Ostrom (2007: 74) reports that "many consumers had equated 'local' with a particular idealized type of farmer or their relationship to a farmer, making such associations as small, independent, or trustworthy."
2. *Nonlocal* is meant to be a value-neutral term. It avoids some of the confusion that would arise if we used "conventional" or "traditional" to describe these products.
3. The pilot case study on Lake Country Cooperative is available online at foodindustry center.umn.edu/Local_Food_Case_Studies.html.
4. Sample questions for supply chain participants are available online at foodindustry center.umn.edu/Local_Food_Case_Studies.html.
5. The recorders' guide is available online at foodindustrycenter.umn.edu/Local_Food _Case_Studies.html.

References

Durham, Catherine A., Robert P. King, and Cathy A. Roheim. 2008. "Consumer Definitions of 'Locally Grown' for Fresh Fruits and Vegetables." Presentation at the Food Distribution Research Society Meeting, Columbus OH, October 11–15.

Hansmann, Henry. 1996. *The Ownership of Enterprise*. Cambridge MA: Belknap Press of Harvard University Press.

Hart, Oliver. 1995. *Firms, Contracts, and Financial Structure*. Oxford: Clarendon Press.

King, Robert P., and Luciano Venturini. 2005. "Demand for Quality Drives Changes in Food Supply Chains." In *New Directions in Global Food Markets*, ed. Anita Regmi and Mark Gehlar, 18–31. Washington DC: U.S. Department of Agriculture, Economic Research Service. www.ers.usda.gov/publications/aib794/aib794d.pdf.

Lyson, Thomas A. 2007. "Civic Agriculture and the North American Food System." In *Remaking the North American Food System: Strategies for Sustainability*, ed. C. Clare Hinrich and Thomas A. Lyson, 19–32. Lincoln: University of Nebraska Press.

Marsden, Terry, Jo Banks, and Gillian Bristow. 2000. "Food Supply Chain Approaches: Exploring Their Role in Rural Development." *Sociologia Ruralis* 40: 424–38.

Ostrom, Marcia. 2007. "Everyday Meanings of 'Local Food': Views from Home and Field." *Community Development* 37: 65–78.

Stevenson, G. W., and Rich Pirog. 2008. "Values-Based Supply Chains: Strategies for Agrifood Enterprises of the Middle." In *Food and the Mid-Level Farm: Renewing an Agriculture of the Middle*, ed. Thomas A. Lyson, G. W. Stevenson, and Rick Welsh, 119–43. Cambridge MA: MIT Press.

Williamson, Oliver E. 1975. *Markets and Hierarchies: Analysis and Antitrust Implications*. New York: Free Press.

———. 1986. *Economic Organization: Firms, Markets and Policy Control*. New York: New York University Press.

Womack, James P., and Daniel T. Jones. 1996. *Lean Thinking*. New York: Simon and Schuster.

Womack, James P., Daniel T. Jones, and Daniel Roos. 1991. *The Machine that Changed the World*. New York: HarperPerennial.

Yin, Robert K. 1994. *Case Study Research: Design and Methods*, 2nd edition. Applied Social Science Research Methods Series, vol. 5. Thousand Oaks CA: Sage Publications.

Part 2

Case Studies on Local Food Supply Chains

3 Apple Case Studies in the Syracuse MSA

Miguel I. Gómez, Edward W. McLaughlin, and Kristen S. Park

Introduction

This case describes the movement of apples through three different marketing channels in Syracuse, New York:

- a supermarket chain (mainstream supply chain),
- a producer who sells at a farmers market (direct market supply chain), and
- a school district that purchases local apples for inclusion in school lunches (intermediated supply chain).

The production area for local food products is defined as the entire state of New York for these case studies. The three case studies describe the nature of the three different supply chains (mainstream, direct, and intermediated) to examine and evaluate differences and similarities regarding supply chain structure, size, and performance. The case studies also discuss prospects for expansion of local apples in each supply chain. The direct and intermediated supply chain cases are used as examples of local supply chains and are compared with the baseline mainstream supply chain case.

The Location: Syracuse MSA

The Syracuse Metropolitan Statistical Area, located in central New York, has a population of approximately 720,000, including the counties of

35

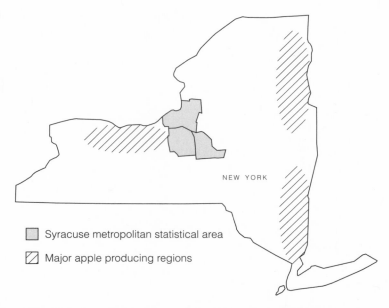

Fig. 1. New York State, Syracuse MSA and major apple producing regions.

Onondaga, Oswego, and Madison (figure 1). It exhibits "average" demographics for upstate New York in terms of ethnicity, education, household composition, and income. It is often used by food manufacturers as a test market for new products due to its demographic representativeness.

Syracuse has six of the top 75 U.S. food retailers, including national chains such as Wal-Mart and Aldi, and four regional supermarket companies, each with five to ten stores operating in the metropolitan area. Syracuse also has a small group of independent supermarkets, a large number of ethnic markets, and a few food market cooperatives.

The Product: Apples

A distinct feature of these apple cases is that New York is both a major apple producer and an important destination market for other apple-producing regions in the United States and elsewhere. New York is the second largest apple-producing state, with the largest concentrations of apple production along Lake Ontario, the Champlain Valley, and the Hudson Valley (figure 1). Table 1 shows that in 2008 New York had about

Table 1. U.S. and New York apple statistics, 2008

SOURCE	VARIABLE	U.S.	NY	NY AS % OF U.S.
1	Bearing age acres, *acres*	350,590	42,000	12.0
1	Yield per acre, *pounds*	27,400	29,800	108.7
1	Utilized production, total, *million lb*	9,676	1,250	12.9
1	Value of utilized production, total $ *millions*	2,187	255	11.7
1	Utilized production, fresh, *million lb*	6,243	520	8.3
1	Value of production, fresh, $ *millions*	1,884	162	8.6
2	Grower price, fresh, $ *per lb (packing house door)*	0.296	0.312	na
2	Retail price, $ *per lb*	1.32	na	na
2	Fresh consumption per capita, *lb*	16.2	na	na

Retail prices are for Red Delicious apples and are adjusted to allow 4 percent for waste and spoilage incurred during marketing. Grower price is for marketing season 2008–2009.

Sources: 1. Noncitrus Fruits and Nuts 2009 Preliminary Summary, usda.mannlib.cornell .edu/usda/nass/NoncFruiNu//2010s/2010/NoncFruiNu-01–22–2010_revision.pdf (accessed 12/8/2011). 2. Fruit and Tree Nut Yearbook 2009, usda.mannlib.cornell.edu /MannUsda/viewDocumentInfo.do?documentID=1377 (accessed 12/8/2011).

42,000 acres planted to apples, which produced 1.25 billion pounds, accounting for almost 13 percent of the national crop (USDA-NASS 2009). Washington is the leading apple producer, growing more than 50 percent of all fresh apples in the country. While New York is the second leading apple producer, its production is significantly smaller than Washington's.

The U.S. average grower price for fresh market apples in 2008 was almost 30 cents per pound, and the average retail price for Red Delicious apples was $1.32 per pound. In the case of fresh apples, the growers received about 22.4 percent of every dollar spent by the end consumer on fresh apples. Because fresh apples, by definition, are not processed and have low shrink relative to other fresh produce items, the grower share of the retail "food dollar" is somewhat higher than that for many other perishable food products.

National apple consumption in 2008 was about 16.2 pounds per capita, making apples the second most popular fresh fruit in the United States after bananas. Organic apple production is very small compared with conventional production, and growing apples organically is difficult in New York due to its climate. While it has ample water and rainfall, this climate is also conducive to high insect and fungal load. Producers have found that in general, pesticides are needed to control apple pests. New York had only 225 acres of certified organic apples in 2007, less than 1 percent of its apple acreage. Washington had 6,681 acres of organic apples in 2007 and California had 3,403 acres (USDA-ERS n.d.a).

Nationally, 67 percent of total apple production is consumed fresh. The remainder is consumed in processed form, such as juice, sauce, cider, and other apple-based processed products. In New York only about half of the crop is sold fresh, with the remainder sold to processors. For the fresh market, as well as the processed market, New York produces more apples than it consumes on a per capita basis. Virtually all the mainstream retail channels in the state purchase fresh apples from New York apple producers, and their distribution centers are easily within a couple of hours away from any New York apple farm. Despite the fact that New York produces an ample supply of apples, not all apples consumed in Syracuse are from New York farms due to seasonality of production, consumer demand for variety, risk diversification among production areas, and supermarket volume requirements.

Table 2. Value of agriculture products sold directly to individuals for human consumption, New York, 2007

DIRECT SALES	U.S.	NY
Number of farms	136,817	5,338
Value ($ *millions*)	$1,211	$77
Average value per farm	$8,853	$14,512
Population	301,290,332	19,429,316
Average value per person	$4.02	$3.99

Sources: 2007 Agricultural Census, agcensus.usda.gov/Publications/2007/Full_Report /Volume_1,_Chapter_2_US_State_Level/st99_2_002_002.pdf (accessed 12/14/2011), and U.S. Census Population estimates, census.gov/popest/index.html (accessed 12/14/2011).

Direct apple sales through farm stands, farmers markets, and U-pick operations are also important sources of revenue for some apple producers in the eastern United States where producers are close to major consumption areas such as urban and suburban markets. Because of the long history of apple production in the eastern United States, and primarily in New York, apple producers are also able to offer many apple varieties, including regional favorites as well as new and even heirloom varieties (Merwin 2008).

In 2007 in New York, a state well known for the number of farm stands and farmers markets, the total value of direct apple sales was $77 million, and direct marketing annual sales per farm were $14,512, almost twice the national average (table 2; USDA-NASS 2007). The average per capita value of agricultural products sold directly to consumers in New York is comparable to the nationwide average (about $4/person). The New York State Department of Agriculture and Markets supports direct marketing activities through various statewide programs and lists approximately 450 farmers markets on its web site.[1]

Mainstream Case: SuperFoods

The focal store in this case is located in Syracuse and belongs to a regional supermarket chain called SuperFoods, which is vertically integrated,

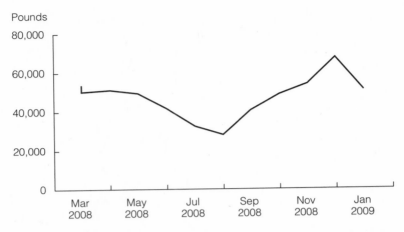

Fig. 2. Apple volume by month in 2008, focal store, Syracuse, New York. Source: Company records.

whereby the distribution center and retail operations are integrated in one company.[2] In 2008 the store sold 1.5 million pounds of apples, with an annual seasonal pattern marked by a peak in volume from September to early December and a low from June to August (figure 2). SuperFoods offers a wide assortment of apples: more than 18 varieties as well as various presentations, including organic, conventional, bulk, and bagged. The supermarket company has a seasonal produce program for smaller local farmers, but the focal store chose not to include apples in the program.

Figure 3 traces the supply chain for apples sold through the focal store. Five apple suppliers provide nearly 100 percent of the apples moving though the supermarket distribution center, and this core group of suppliers has had commercial relationships with the supermarket chain for many years. The suppliers come from two distinct geographic regions: Washington and New York. The supermarket chain does not import apples directly but occasionally buys imported apples from its main suppliers in Washington and New York, which in those cases serve as importers.

Production

Four of the five sources of supply for apples are vertically integrated grower-packer-shippers (GPSs). The production, grading, packaging,

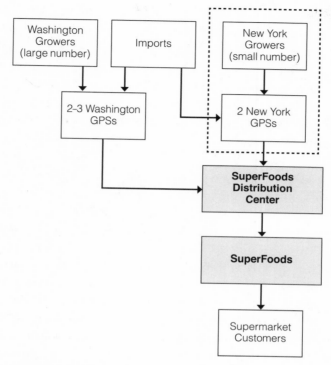

Fig. 3. Mainstream supply chain, SuperFoods.

storage, and selling functions are all performed internally by these firms. Two of these GPS firms are from New York (GPS1 and GPS2), and two are from Washington (GPS3 and GPS4). All these GPSs buy apples from other growers to supplement their own production. The fifth source of supply is a broker from Washington. About 75 percent of apples in the focal store are sourced from New York and 25 percent from Washington.

NEW YORK SUPPLIERS: The two New York suppliers operate similarly, so the discussion here focuses primarily on GPS1. GPS1 supplies about 35 percent of SuperFoods' apples, has about 2,500 acres in production, and offers 23 distinct apple varieties to its customers; about 75 percent of its production grades out at the highest quality, "Extra-Fancy." GPS1 employs nearly 290 seasonal workers in the orchards and 40 workers in the packing shed, which operates year-round. It also owns six trailer trucks with which it delivers within 300 miles of its packing/storage facility. It uses a trucking broker for longer hauls. GPS1 has begun experimenting with organic production and makes use of integrated pest management production methods developed by Cornell University.

GPS1 also supplies apples merchandised as "local" during a 12-week harvest period from early September through late November. These apples are also delivered to the SuperFoods distribution center in four-pound tote bags priced at $0.89/lb, which is lower than apples merchandised in the standard apple display in the store.

About 80 percent of the sales from GPS1 are sourced from its own farms, with the remaining 20 percent sourced from 20–25 independent growers within New York and from importers. Prices paid to these independent growers vary by season, storage time and method, package, and quality.

GPS1 sells directly to approximately 20 retail accounts, which comprise 75 percent of its sales; 25 percent of sales are to processors. GPS1 has no food service accounts and considers this a weak market option for apples, partly due to the perceived domination of the national food service market by Washington apple suppliers. Sales volume for GPS1 is seasonal, peaking in September–November and steadily declining through May–July, when other summer fruit becomes available and demand for apples falls. Prices are highest during fall harvest and begin to drop after Thanksgiving. GPS1 views its greatest competition in the future coming from the Southern Hemisphere.

NEW YORK GPSS AS "LOCAL SUPPLIERS": GPS1 and GPS2 are both located in New York and are both family owned and operated. Thus many observers might consider them to be "local producers," because of other connotations brought up by the term *local*. However, these two GPSs do not share the characteristics that many consumers envision as local. They are quite large relative to most other New York fruit growers, with sales extending down the East Coast and abroad. They do not engage in sales directly to consumers through either farmers markets or farm stands. Also, given their volume, the logistics of direct-store delivery are generally too challenging to pay off. It would be uneconomical for these GPSs to attempt to send a shipment of apples by their own trucks to each supermarket in each of their accounts. In almost all respects, the New York GPSs follow procurement and selling practices similar to those of their counterparts in Washington and thus are more like the West Coast suppliers than some smaller New York growers that sell in farmers markets or farm stands.

WASHINGTON SUPPLIERS: Although the standard procurement, operating, and selling procedures for the Washington suppliers are generally similar to those of the New York suppliers, some differences are noted here. GPS3 and GPS4 are integrated grower-packer-shippers, similar to GPS1 and GPS2 in New York. The fifth supplier is a broker from Washington that facilitates apple sales between various grower-packers in Washington and SuperFoods, and earns a commission on such sales, but does not take title to the apples. About 25 percent of apples arriving at the SuperFoods distribution center are sourced from these three Washington suppliers. The Washington GPS3 and GPS4 are much larger than the New York counterparts and, in fact, figure among the largest fruit and apple suppliers in the United States. GPS3, for example, sells apples from approximately 3,100 acres, which amounts to roughly 8 million boxes from 70 different growers. Although it sometimes needs to buy packed apples or other produce to meet demand, GPS3 packs nearly all—85 percent—of the apples it sells. SuperFoods represents an important customer for the Washington suppliers but, in contrast to the situation for the GPSs from New York, is not estimated to be in the top ten customers by volume for any of the three Washington suppliers.

Distribution and Marketing

All the apples sold at the focal store arrive first at SuperFoods' distribution center. From there the category manager directs the pricing, merchandising, sourcing, and product assortment for the apple category and manages the strategic direction of the category. A buyer, working under the direction of the category manager, gathers individual store orders, places the aggregated orders with suppliers, tracks inventories, and deals directly with invoicing. The store produce manager places orders daily to the buyer and distribution center for next day delivery.

The apple buyer also tracks apple inventories at the distribution center and places orders with suppliers to maintain an optimal level of inventory in the distribution center. A primary objective of the category manager and buyer is to ensure that the distribution center has an adequate supply to avoid running out of apples in the stores. At the same time, they aim to keep high turnover to maintain freshness and quality and to minimize inventory costs. Adjusting inventory levels to achieve this key balance is critical. Consequently the buyer generally places orders five days a week to various suppliers.

DEMAND FORECASTS: The SuperFoods buyer uses a simple spreadsheet model to forecast demand from the stores. The forecast assumes the volume of apples that will be ordered by the focal store on a given day is equal to the volume ordered the same day the previous week. The buyer can override the forecast manually to accommodate unusual demand conditions like seasonal patterns of consumption or holidays.

ORDERING AND TRANSPORTATION: The apple buyer places electronic orders every day for delivery to the distribution center, which usually takes two days for New York apples and 5–6 days for Washington apples. Within this time, there may be adjustments to the order up to the time the trucks are loaded by the GPS.

GPS1 and GPS2 are both located within about 100 miles from the SuperFoods distribution center. The normal order batch is a full tractor-trailer load, 40,000 pounds, delivered to the retail distribution center in the GPS-owned trucks. The GPS charges the retail headquarters the cost of transportation from the farm to the distribution center, roughly $500 per delivery.

SuperFoods places electronic orders with Washington suppliers twice per week, with expected delivery time typically 7–8 days from time of order placement to arrival at the SuperFoods distribution center. This time frame allows 1–2 days to assemble the order and 5–6 days for transport, considering suppliers are located approximately 2,600 miles away from the distribution center. Free-on-board (FOB) apple prices—that is, prices at shipping point—are generally 10 percent to 30 percent lower in Washington than in New York. According to SuperFoods' Washington suppliers, these price disparities exist for three reasons: (1) yield per acre is more than 20 percent higher in Washington (22 tons/acre) than in New York (18 tons/acre); (2) horticultural costs (e.g., sprays) are lower in Washington due to more favorable climate; and (3) Washington suppliers are so much larger that they can spread fixed costs over far more volume, including fruit sales other than apples. Most apple grower cost studies estimate that approximately half of the FOB costs are production-related and half handling-related. Thus although large variances occur, the return to the grower(s) may average roughly half of the shipper's FOB price.

By the time the apples reach SuperFoods much of this production cost advantage is offset by the cost of long-distance transport, which can vary between $4,000 and $8,000 per truck load of apples from Washington to the SuperFoods distribution center. The large variance is caused by the changing seasonal demand for trucks; it costs much more to hire a truck during the peak summer demand. As a result, retailer cost per box for Washington apples delivered to the SuperFoods distribution center is often actually marginally higher than for New York apples.

Whether from New York or Washington, suppliers deliver to the supermarket distribution center and charge the supermarket chain for the cost of transportation from the packing houses to the distribution center. The two New York GPSs use their own trucks to deliver to Super-Foods, while the suppliers in Washington generally rely on third-party owner-operated trucks.

Each order is invoiced separately, and every load is inspected at the distribution center upon delivery. Outright rejections are rare for apples; less than 1 percent of loads are turned away. In some cases a detected problem may only be in a single lot number. Once the load is accepted, the invoice is signed and transmitted to retail headquarters, and the GPS receives payment from the retailer within 25 days.

The distribution center, located within 100 miles of the focal store, makes deliveries six times per week to the store in its corporately owned trucks. The distribution center sends full tractor-trailers to stores with apples sharing space with other produce items and occasionally other fresh foods. Three factors facilitate coordination along the supply chain. First, under proper conditions, apples can be kept in storage longer than most produce items. As a result, suppliers know their annual inventories quite precisely once harvest is complete, unlike for many fresh produce items, for which "inventories" are typically in constant flux due to weather, pests, and other pre-harvest conditions. Second, developments in controlled atmosphere storage and in ethylene treatment (under the brand name SmartFresh) of apples during storage have markedly improved storage life and quality. Third, the supermarket chain employs an electronic data interchange system that substantially facilitates placing orders to suppliers, monitoring product inventories, and receiving orders from the focal store. All vendors and the distribution center are responsible for maintaining this database, which includes items available, projected supplier prices, and current inventory levels. Although this system contributes to efficiency and coordination, there is also continuous near daily communication between the buyer and the five suppliers.

GPS1's management meets with the SuperFoods headquarters management about six times per year and hosts an annual farm visit by more than 100 people from SuperFoods, including both store and headquarters personnel. Such activities can build a certain "social capital" related to supermarket procurement methods.

When asked, the category manager stated that the criteria the retail company values most from its apple suppliers are service levels, freshness, and inventory control. It is perhaps significant to note that the criteria GPS1 mentioned as being of most value to the retailer were "consistency and innovation." In neither case was "quality" listed, although freshness is certainly one dimension of quality.

MARKETING AND RETAILING: The focal store is one of the largest in the retail chain. The produce department has 32 employees, who rotate across the different categories so that all produce personnel know procedures for all products. The apple category is one of the largest produce categories, accounting for between 10 percent and 15 percent of produce sales—nearly twice the industry average of about 6 percent.

Both as a positioning strategy and to help control inventory and "shrink" (loss from waste, damage, and theft), the retailer employs an "every day low price" (EDLP) strategy—whereby prices remain almost constant week-to-week with few promotional discounts. In addition, the focal store employs a limited number of prices across apple varieties and presentations. Such an EDLP strategy, together with only a few price differences among apple varieties, allows the supermarket chain to streamline ordering and to simplify inventory control at the store and distribution center.

The focal store offers an astonishing assortment of apples throughout the year. The ordering system lists 129 different stock keeping units

(SKUs). This includes differences not just in variety but in pack size and growing condition—organic versus "conventional." Some 45–50 SKUs are consistently available for 18 varieties, throughout the year and in various merchandising combinations—organic, conventional, bulk, and bags (3, 5, and 8 lb). In 2008 bulk apples sales accounted for approximately 53 percent of the store's apple sales, and the top 10 apple varieties, which were sold in bulk, accounted for 72 percent of the bulk apple movement in the focal store. When considering bagged sales, the three-pound bag was the unit of preference by far, accounting for nearly 70 percent of all bagged apples sales in 2008.

With limited in-store cold storage, the produce manager generally orders apples from the distribution center six times per week, with expected store delivery the following day. In principle, the focal store is also permitted to order from a list of approved direct suppliers, but in fact, the store places orders for paper tote bags of apples from the distribution center (supplied by GPS1) during the harvest season.

No food safety issues associated with apples delivered to the retail distribution center have been reported. The five major apple suppliers delivering to the retail distribution center carry food safety certifications. GPS1, for instance, has the GLOBALG.A.P. certification, including an annual inspection and sporadic unannounced inspections. GPS3 in Washington is certified with the Safe Quality Food (SQF) program, which is run by the Food Marketing Institute. Both programs include safe production practices (SQF 1000) as well as certifying the packing house facilities (SQF 2000).

PRICES ALONG THE SUPPLY CHAIN: The retail price of apples sold in bulk in the focal store in 2008 and supplied by GPS3 (from Washington) was $1.89/lb. The retail margin was $1.00/lb (or 53 percent of retail price). The transportation cost from the packing house to the distribution center was $0.23/lb (12 percent of the retail price), and the FOB price at the packing house was $0.66/lb (35 percent of the retail price).

For New York apples, the retail price of a typical bulk apple variety from GPS1 sold in the SuperFoods store was $1.50/lb in 2008. The retail margin was $0.76/lb (51 percent of retail price), lower than the margin for a Washington grower by about $0.24/lb. GPS1 delivers apples both with its own trucks and by contracting with third parties. In 2008 the transportation cost was about $0.03/lb (2 percent of the retail price). GPS1 also provides apples in various bag sizes to SuperFoods. The retail blended price of all three bag sizes was about $1.00/lb in 2008, thus producing a retail margin of $0.37/lb (37 percent), transport cost of $0.03/lb (2 percent), and FOB price to GPS1 of approximately $0.60/lb (60 percent). The difference in transportation costs of about $0.20/lb represents a cost advantage of New York apples over those originating in Washington.

Food Miles and Transportation Fuel Use

Apples in the mainstream supply channel are transported from farms to packing house(s), from packing house(s) to the retail distribution center, and from the distribution center to the focal Syracuse store (see figure 3).

WASHINGTON STATE APPLES (GPS3): The average distance from the farm gate to the packing house is 150 miles in trucks with fuel efficiency of 12 miles per gallon (mpg).[3] The trucks, with capacity to transport 10,000 pounds, travel from the packing house to the farms and back to the packing house, for an average distance of 300 miles. Therefore the average fuel use per 100 pounds from the farm gate to the packing house is 0.25 gallons. GPS3 delivers apples to the SuperFoods distribution center with trucks that get 6 mpg and have capacity to transport about 40,000 pounds of apples. Apples from the Washington State travel 2,600 miles and therefore use 1.08 gallons of fuel per 100 pounds (one-way fuel use is relevant since backhaul loads are employed). In turn, apples are hauled from the distribution center to the focal store. The two-way 200-mile truck trip uses 33.3 gallons of fuel, and the average delivery to the focal store is about 2,000 pounds. Of course, the truck carries other goods as well—apples occupy only about 5 percent of the truck capacity. Therefore the fuel use per 100 pounds of apples equals 0.08 gallons. Adding fuel use across the three segments of the supply chain, total transportation fuel use is 1.41 gallons per 100 pounds of apples over a total distance of 3,100 food miles.

NEW YORK STATE APPLES (GPS1): GPS1 integrates grower, packer, and shipper operations. The average distance from the farm gate to the packing house is 25 miles in trucks with fuel efficiency of about 12 mpg. Trucks have a capacity to transport 10,000 pounds. They pick up the product at the farm gate and bring it to the packing house for an average roundtrip of 50 miles. Therefore the average fuel use per 100 pounds is 0.04 gallons from the farm gate to the packing house. GPS1 is located about 100 miles from the distribution center. It usually transports a full tractor-trailer load (about 1,000 boxes, each weighing 40 pounds) to the retail distribution center. The tractor-trailer returns empty to the packing house, yielding 0.08 gallons per 100 pounds for the two-way trip. Once in the retail distribution center, apples from New York are handled the same way as apples from Washington. They are put in storage and hauled from the distribution center to the focal store in Syracuse in a company-owned tractor-trailer. The 200-mile roundtrip uses 33.3 gallons of fuel and the average delivery to the focal store is 2,000 pounds. Again, given that apples only occupy 5 percent of truck capacity, fuel use per 100 pounds of apples equals 0.08 gallons. Adding across the supply chain indicates that total transportation fuel use is 0.20 gallons per 100 pounds, substantially less than the per pound fuel use of apples coming from Washington.

Community and Economic Linkages

One of the primary functions of a supermarket is to offer a space where consumers can make all their purchases in a single trip. Consumers benefit from less time overall spent shopping, fewer trips and miles to other food outlets, high food safety standards, and wide product assortment. In addition, the focal SuperFoods store provides substantial support to the local community, evidenced by multiple activities from providing supplies for the local food bank to funding scholarship programs for employees.

SuperFoods' apple supply chain has important impacts on employment and local economic development. The focal store employs more than 500 people on a full-time or part-time basis, 32 of whom work in the produce department (of course working with more than apples only), and nearly all of them reside in the local community.

This apple supply chain also generates substantial employment along the supply chain, specifically in such activities as growing, handling, shipping, storage, and transportation. The New York supplier GPS1 generates nearly 400 jobs in upstate New York and buys apples from 20–25 medium-sized and small farms in the region, each generating additional jobs on the farm.

Prospects for Expansion

SuperFoods has a seasonal local program for produce, and the focal store is allowed to purchase produce for direct store delivery from a list of approved vendors. Store management points out that carrying local produce is in direct response to shopper requests: if local produce is not present, shoppers ask why it is not available, even when items are not in season. However, apples are not generally included in the store's direct-delivered produce program. The store manager reports that in many ways, he prefers to receive tote bags packed by GPS1 during the harvest season from the SuperFoods distribution center because the ordering process is easier to control, apples come in handy cases (not bins), and more varieties are available than from the approved list of local direct-delivery apple growers. While tote sales are strong and customers perceive them to be "local," the retail gross margin applied to these apples is considerably below the produce department average and far below the margin of other apples. SuperFoods has little incentive to increase sale of tote apples beyond the minimum required to produce a certain "local" image.

This case identifies several barriers for expansion of local apples delivered directly to the store (i.e., bypassing the distribution center), particularly for small apple growers. First, smaller local growers are not likely to have the scale to meet the single-store volume requirements of SuperFoods, even during the local apple season. Second, some smaller local growers believe they can get higher prices or higher returns from selling through direct channels, such as farmers markets, rather than to supermarkets like SuperFoods. Third, according to SuperFoods, consumer demand requires that local apples have nearly the same quality and appearance as other apples in the store. Therefore smaller local growers have to make at least minimal postharvest investments in refrigeration, packing facilities, and carton quality. Economies of scale associated with these investments may hinder the ability of small local growers to increase the share of local apples in the supermarket. Fourth, local apples can decrease store operational efficiencies, as small local growers often deliver in small quantities on a frequent basis, implying potential increased labor and administrative costs to the retail operation. Finally, small local growers may have a limited ability to offer the assortment of product required by the focal store.

Food safety assurance may be another barrier to local apples delivered directly to the store. SuperFoods requires equal food safety standards from all its produce suppliers regardless of farm size. Meeting certification standards that have been developed primarily for large producers has been challenging for small producers. Consequently, SuperFoods has worked with smaller growers to help them meet such standards. The company, in collaboration with major universities and with the USDA, has been training local small growers how to be in compliance with the USDA Good Agricultural Practices (GAP) and has been helping them obtain GAP certification since 2005. Although apples are not considered a high-risk commodity, SuperFoods is requiring that all local growers delivering directly to stores obtain third-party GAP certification by the end of 2009. Local growers who are not willing or not able to obtain GAP certification will not be allowed to supply the company's stores.

Key Lessons

This case study provides several lessons regarding the mainstream supermarket channel for apples and its relation with local food supply systems.

First, the supermarket channel may be well positioned to play a critical role in increasing the share of local foods in total food sales. In fact, New York suppliers GPS1 and GPS2 are large local suppliers that enjoy a large market share in SuperFoods' apple category. The supermarket channel is by far the largest food distributor, and consumers shop for food in this channel in far greater numbers than in any other. Nevertheless, the barriers to local produce grown by small farmers and delivered directly to the store discussed in this case are very hard to overcome. Consequently it is probable that direct store deliveries of produce will remain a small share of total food sales in the foreseeable future.

Second, the case shows that the revenue shared by the grower-packer-shippers decreases with distance to the destination market. In the case of bulk apples in Syracuse, the value shares for the New York and Washington suppliers are 47 percent and 35 percent of the retail price, respectively. Nevertheless, the differences in dollars are much smaller given that the retail price of apples is greater for bulk apples from Washington versus New York: $1.89/lb and $1.50/lb, respectively.

Third, this case indicates that fuel utilization per 100 pounds for apples coming from the West Coast is significantly higher than for apples sourced in New York, even though West Coast production costs and FOB prices are lower. These results indicate that the two large New York suppliers may be the most efficient in terms of fuel use, at least when serving local New York markets.

Fourth, this case shows the strengths of the supermarket channel. SuperFoods has developed an efficient supply chain able to provide the end consumer with year-round availability of a wide variety of products at low prices. There is a close relationship between channel members that facilitates information sharing. This leads to improved channel coordination. In addition, the supermarket offers certain conveniences to consumers, such as less time spent shopping and fewer trips and miles traveled for food purchases. Therefore it is likely to be quite difficult for direct distribution channels of local foods to compete with the products and services provided by the supermarket channel.

Direct Market Case: Jim Smith Farm

The direct market for this case study is the Central New York (CNY) Farmers Market in Syracuse. The CNY Regional Market Authority owns and administers a farmers market and a wholesale terminal market, both located on the same property. However, the two units are completely independent in their operations. The Market Authority completed an $8.4 million upgrade to the combined site in 2001. The farmers market serves as a historic landmark for the city, and future plans call for a total revitalization of each of the 18 buildings located at the market.

This farmers market started operations in 1938, and it is extremely popular among citizens of Central New York. Presently the farmers market hosts about 300 vendors. On a typical Saturday the farmers market receives over 10,000 visitors who browse and buy a wide variety of fruits, vegetables, and flowers.

The focal vendor for this case study, which we call Jim Smith Farm, is an independently owned and operated family farm.[4] The farm has had a stall in the CNY Farmers Market for 10 years, and its owner recently became a member of the market's Board of Directors. Smith has 115 acres, of which 90 acres are farmed. Although its major crop is apples, the farm produces a wide variety of fruits and vegetables, also including peaches, sweet cherries, raspberries, plums, pears, asparagus, squash, cucumbers, and beans. Smith has average annual revenues of about $250,000. Having a diversified farm operation helps the farm participate in direct marketing initiatives such as farmers markets, spread out the cost of labor, and extend the production season. Apples are by far the most important product for the farm and are produced on about 50 acres (more than half the farmed acreage). Smith grows such popular varieties as Gala, McIntosh, Macoun, Empire, Golden Delicious, Red Delicious, and Fuji. In addition, it produces certain less common and heirloom varieties such as Yellow Transparent, Ginger Gold, Jonagold, Northern Spy, Rome, Honeycrisp, CandyCrisp, Ambrosia, Autumn Crisp, and Zestar, although these represent minor volumes. The farm's owner also operates a second business that sells, services, and designs drip irrigation for greenhouses, nurseries, and vineyards.

Figure 4 traces the supply chain for apples sold by Smith in the farmers market. The farm employs two primary distribution channels: direct marketing, mostly through farmers markets, and wholesale to a packer-shipper company. About 5 percent of the total apple production is sold through farmers markets and about 95 percent is sold to the packer-shipper.

In addition to selling apples in Syracuse's CNY Farmers Market, Smith sells in three other farmers markets in Central New York—Auburn, Seneca Falls, and Syracuse downtown. In total the farm is present for eight market days per week in these four markets. The largest share of Smith's direct sales is made through the CNY Farmers Market. On certain occasions Smith sells apples to farm stands nearby but in very small quantities.

Production

Smith produces about 40,000 bushels of apples per year from about 50 acres of apple orchards. The farm offers roughly 20 different apple varieties, including both popular and less common varieties. All apples are conventionally produced, and all apples sold by the farm are sourced from its own production.

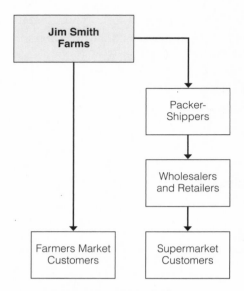

Fig. 4. Direct market supply chain, Jim Smith Farm.

Smith is devoted full-time to the business, which includes the farming operation and the irrigation supply business unit. He divides his time equally between these two activities. Smith employs one part-time production employee year-round. During the apple season, labor requirements increase dramatically and 10 seasonal employees are hired from April to October. Eight of these workers are hired under a federal program for temporary or seasonal immigrant labor called H-2A certification. The remaining two are hired from the local labor market. In addition a retired farmer from a nearby farm works three-quarter time with the equipment-related activities during the crop season. Smith owns two box-vans with which it delivers to all its direct channels.

Apples are picked in 20-bushel bins. A typical bushel holds 36 pounds of apples, so each bin contains approximately 720 pounds. Smith plans harvesting activities for the direct marketing channel and for sale to the packer-shipper on a weekly basis to minimize cold storage requirements. Apples sold to the packer-shipper are transported in box-vans to the packer-shipper immediately after picking on a daily basis during harvest season. Apples for the farmers markets, on the other hand, are kept in coolers on the farm or in a nearby commercial cooler located 10 miles away. These apples in the commercial cooler are kept in storage for about two days on average at a cost of \$15/bin/day for regular storage and \$35/bin/day for controlled atmosphere storage.

Although the farm does not pack according to USDA grade or size standards, the best apples are generally selected for the farmers markets. Heirloom and less common varieties are generally destined for the direct marketing channel, while the most common varieties (e.g., Red Delicious, Golden Delicious) are sold to the packer-shipper.

Apples for the farmers market are packed in two-, four-, and eight-quart tote bags and also in half bushels. Apples for the packer-shipper are delivered in 20-bushel bins. Sales volume for Smith is seasonal, peaking in September through mid-November with the apple harvest and steadily declining through December, when the farm stops selling apples. It resumes selling activities in late April during asparagus season. Smith sees its greatest opportunities in the direct marketing channels, which have been growing at an annual rate of 50 percent during each of the past three years.

Marketing and Distribution

Although apples are its primary crop, Smith sells a wide variety of fruit and vegetables grown on the farm. Depending on the time of the year, apples represent between one-half and one-third of sales in the farmers market. Smith sells on Tuesdays, Thursdays, and Saturdays in Syracuse's CNY Farmers Market, from late April through Christmas. Its product assortment is typical of other apple vendors attending the farmers market.

Smith staffs the farmers market operation with three employees: the owner and two other family members. They are responsible for loading the products in the van, setting up the stand at the farmers market, selling the product, and closing the stall at the end of the day. On most occasions Smith sells all the products brought to the market, so the van generally goes back to the farm empty. The farm sells a wide assortment of apples through the season, offering about 20 different varieties in various merchandising presentations including tote bags of two, four, and eight quarts. In 2009 these were priced at \$2.00, \$3.00, and \$5.00 per unit, respectively. Smith also sells apples in half bushels for \$9.00 each. The majority of sales are two- and four-quart tote bags.

All farmers market sales are in cash, except for sales to people in the federal food assistance programs, such as the Women, Infants, and Children Program and the Supplemental Nutrition Assistance Program (SNAP). In these cases Smith collects the electronic benefit transfer (EBT) tokens used by the farmers market management to administer the payments. He then exchanges the tokens for their value as needed from the market office.

When asked about the economic benefits of participating in the farmers market, Smith estimated that prices from the farmers market operation average $360/bin, which is much higher than the average price of about $200/bin obtained by selling to the packer-shipper. This translates into a return to Smith of $0.50/lb from the farmers market versus $0.28/lb for apples sold to the packer-shipper. On a typical Saturday Smith may sell 50 to 60 bushels of apples, yielding revenue of $900 to $1,080 per market day. During weekdays, apple sales are about half those on a Saturday. The owner is highly satisfied with his participation in the CNY Farmers Market.

Jim Smith Farm, located about 60 miles from the regional farmers market, assumes all costs associated with transportation from the farm to the market. The family loads the van and leaves for the market by about 5:30 a.m. on market days. The van is loaded full with fruits and vegetables. The price received by Smith net of marketing expenses is $0.40/lb or 80 percent of his direct market retail price. This is substantially higher than the average apple grower price, $0.26/lb, as estimated by USDA Agricultural Marketing Service.[5]

According to Smith, market visitors look forward to buying apples in the farmers markets and often ask for specific varieties that may not be available in other retail outlets. When asked about the attributes of his farm's apples most valued by consumers, Smith indicated that the most important factor is consumers' desire to buy directly from the grower. Other reasons include the lower price of apples relative to retail store prices, the perception of freshness as evidenced by taste, and the variety offered in the market. Smith believes that he could sell more apples if he could find more retail space at the farmers market. However, expansion is difficult because all stalls are currently full and vendors rarely leave.

In addition to capturing a large portion of the value in the supply chain, Smith greatly enjoys the interaction with end consumers. This channel has also proven to be an effective way to involve his family in the farm operations. Smith is a member of the farmers market Board of Directors and he describes his relationship with the market manager as excellent and one of mutual support. The greatest challenge for vendors and management alike is to grow the market while keeping the same sense of localness among vendors and consumers.

Apples are not considered a high-risk commodity in terms of food safety. Nevertheless the packer-shipper encourages, but does not require, Smith to be GAP certified. It is possible that more pressure to become certified will come in the future. The farmers market manager stated that he has not witnessed any reported food safety incident. The New York State Department of Agriculture and Markets regulates all food retail establishments, including direct marketing businesses, which are required to comply with certain New York food sanitation requirements. Nevertheless, farmers markets are not required to meet the strict sanitary guidelines required by regular retail food stores or food processing establishments. Vendors in farmers markets that sell fresh whole (uncut) fruits and vegetables are not normally subject to inspections unless the sanitary authorities receive complaints from consumers.

The Farmers Market

The CNY Farmers Market is located on the north side of Syracuse. It operates year-round on Saturdays and Sundays, is open on Thursdays from May through mid-November, and operates on Tuesdays as well from May to the end of September. This farmers market, as noted, has been in operation since 1938, but it has expanded in recent years. More than 300 vendors attend the market on a typical weekend in the summer (with about half the number during weekdays). The average rate of occupancy in the summer is high, nearly 98 percent. The number of stalls roughly doubled from 2002 to 2009. At the same time, this market has become an important food "retailer" among the Syracuse downtown community. According to the market manager, approximately 10,000 people visit the CNY Farmers Market each Saturday from May to October.

The market accepts six types of vendors, three of which sell produce. The three types of produce vendors include farmers selling products exclusively from their own farms, New York resellers that procure all products from within state boundaries, and out-of-state dealers that are allowed to sell products procured from other states. Each vendor-type has differential fees in its leasing contracts. By far, most vendors of fruits and vegetables belong to the first and second categories. Farmers pay $574 per stall for a season contract (May to October) but have the flexibility to lease stalls on a daily basis at a rate of $50 per day. Resellers pay fees that are more than twice those paid by farmer-vendors.

There are 12 apple vendors in the CNY Farmers Market. About half of these vendors are farmers and half are resellers (selling in-state and out-of-state products). Apples make up between 5 and 10 percent of the total sales in the farmers market. Most apples sold in the market are not organic. Vendors offer a wide variety of apples, from popular and commonly consumed apples that are also found in a supermarket to heirloom varieties only available in certain parts of the Northeast.

Most vendors sell apples at the same prices and in the same presentations, and these prices change little through the apple season. In that sense apple vendors appear to be price takers in the market (i.e., they must accept the prevailing prices in the market). On some occasions, vendors who bring a unique apple variety are able to set higher prices compared to those for other apples commonly available in the market. The CNY Farmers Market is the retail outlet that offers the lowest prices for apples among all retail outlets monitored in the Syracuse metropolitan area for this study. This is noteworthy because farmers markets are often perceived to have the highest prices in a given market area.

The market management developed a *Vendor Handbook*, which sets clear rules for market participants, defines the various vendor categories, establishes fees, and explains the permit and license requirements for participation in the market (Central New York Regional Market Authority 2010). The Board of Directors meets once a month to assess market performance and address issues that may emerge through the market season. The CNY Farmers Market supports vendors by taking an active role in marketing activities, advertising on radio and television, in magazines, and occasionally in local newspapers. The market manager also works with the Syracuse Office of Tourism to promote the market among city visitors. The market cannot have its own web site because it belongs to the State Market Authority.

Food Miles and Fuel Use

Apples sold in the farmers market are transported from the farm to downtown Syracuse. Various fruits and vegetables, including apples, are transported in box-vans from the farm gate to the CNY Farmers Market (61 miles) with a fuel efficiency of 20 mpg. The box-van travels back to the farm the same day. Apples occupy about half of the van capacity during the apple season. The box-van typically carries 1,980 pounds of apples on a market day. Therefore, the fuel use per 100 pounds of apples equals 0.16 gallons.

Community and Economic Linkages

The dynamics of the farmers market supply chain result in diverse economic impacts on the economy where the farm is located, on the farm itself, on consumers, and on downtown Syracuse. One of the most important benefits of this supply chain lies in its impacts on the local economy. Upstate New York was once a prosperous industrial region. With industry declining in the Northeast, upstate New York entered a period of extended economic recession as jobs were lost to other regions and countries. It has been argued that value added agriculture could play a key role in the economic sustainability of the region. The focal farm has increased and diversified its production to sell in the farmers market. Today Smith generates about twelve jobs for the local economy during the extended season in which vegetables and fruits, including apples, are planted, harvested, and delivered to the market. Smith starts selling products in the farmers market from early in April (when the asparagus harvest starts) until the end of the year. In addition, the farm often purchases certain services (e.g., apple storage and refrigeration) that are offered by other businesses in the county, and it offers an additional source of income to a retired farmer in the village.

Smith's participation in the farmers market has also impacted employment in marketing activities. Smith participates in farmers markets eight days a week (it sells in more than one market on a given day). Two or three family members are responsible for distribution and retailing activities for each farmers market. The vendor obtains a high level of satisfaction from the interaction with customers and from a sense of having control over the supply chain. Being closer to the end consumers helps Smith make better decisions regarding what products to grow, when, and in what volumes.

Smith and the market management believe that customers benefit from the local supply chain in a number of ways. First, the prices of fruit and vegetables offered at the farmers market are the lowest among retail outlets observed in this study in downtown Syracuse. This is especially important for the communities in downtown Syracuse with depressed income levels. Second, in addition to paying lower prices, consumers appreciate meeting the farmers who grow the fruit and vegetables they consume, and consumers express satisfaction with the sense of freshness of the products they buy. Finally, when asked about the seasonal availability of products, the market managers stressed that having a flexible policy regarding vendor eligibility (e.g., allowing resellers of crops produced in-state and out-of state, but charging higher fees for these types of vendors) helps the retail operation to have a wide variety of products for a good part of the year. Consumers and producers appear to be willing to have marketing of certain nonlocal products in order to have an extended season.

The farmers market also has important spillovers in downtown Syracuse. A recent study by the Farmers' Market Federation of New York (FMFNY) reports that consumers not only spend an average of $10.00 per visit in the Syracuse farmers market on fruit and vegetables but also spend an additional $9.00 in other business in the downtown area during their shopping trip to the farmers market (Farmers' Market Federation of New York 2006). The market manager affirmed that local businesses in downtown Syracuse increase their sales substantially on market days. In addition, the FMFNY study notes that the farmers market has become a tourist destination in downtown Syracuse because it offers a sample of the local culture, from a wide array of products to the diversity of vendors. An additional impact of the farmers market is its contribution to increasing the availability of fresh fruits and vegetables in downtown Syracuse, where most food retail outlets are convenience stores that offer a very limited assortment of fresh produce. Thus the farmers market may contribute to healthier diets of downtown communities. Finally, the farmers market encourages several state and federal food assistance programs aimed at increasing the consumption of fruits and vegetables (e.g., New York's Health Bucks Program).

Prospects for Expansion

This case identifies several factors that facilitate the expansion of apple sales in farmers markets. First, Syracuse is located in a prominent apple-producing region. Therefore the product supply is already there, and the farmers market becomes just an additional distribution channel available to apple growers in diversifying their distribution strategies. Second, the large range of apple varieties with different qualities makes apples an attractive item that can draw consumers to the farmers markets. This is due to New York's production of a wide variety of heirloom and local variety apples, generally unavailable in conventional channels. Third, the farmers market is located in downtown Syracuse in a community where income is lower than the metropolitan area average. According to the market manager, this distribution channel offers the lowest prices for apples among all local alternatives. These three factors, and the fact that the farmers market is open year-round and few supermarkets exist in the downtown area, confer a certain competitive advantage to the farmers market. Fourth, the increased presence in the farmers market of state food assistance programs, like the Farmers Market Health Bucks, and federal food assistance programs like SNAP (delivered through EBT tokens), serve as "pull strategies" to increase the demand for fruits and vegetables, including apples, in the market. This is likely to increase in the future to the degree that various state and federal programs continue to support locally grown produce.

This case also identifies barriers to expansion of apples in farmers markets. First, the ability of Smith to increase its retail space is quite limited. The owner mentioned that he bought a nearby apple farm principally to make sure he could use the retail space of these farmers in the market. The market management is expanding the facilities, but it is very hard to get adjacent stalls, particularly when the location in the market is very good. Second, access to reliable agricultural labor is a critical production constraint for Smith. He requires more labor because he has diversified farm and orchard production to offer a wide assortment of fruits and vegetables in the farmers market. It is not always easy to participate in the H-2A program, and locating suitable labor is always a concern for the farm. Finally, the sales volume of the farmers market is increasing, but it is still a very small portion (5 percent) of Smith's total apple sales.

Key Lessons

This case study provides several lessons regarding the farmers market channel for apples and its relation with the overall apple production and distribution system.

First, the focal farmers market has unique characteristics that contribute to the expansion of local food supply chains. Specifically, this market is open year-round; has flexible rules for vendor eligibility, allowing reselling of in-state and out-of-state fruit and vegetables; is located in a densely populated area where supermarket chains are practically nonexistent; and offers the lowest prices among all retail outlets in the downtown area. Indeed, in this sense, the director of the FMFNY affirmed that the CNY Regional Farmers Market was a very different market than most "typical" farmers markets in the state.

Second, the case suggests that vendor participation in farmers markets impacts farming operations. This case shows that the farmer has diversified his farm's product offerings in order to participate in direct supply chains. The case also suggests that the motivation to participate in farmers markets goes beyond increased profits and extends to vendor satisfaction through the development of entrepreneurial and marketing skills and to the social interaction with customers.

Third, this case highlights the potential of the farmers market for the delivery of state and federal food assistance programs. Farmers markets can be important in delivering programs encouraging citizens to consume more fruit and vegetables, as is the case of New York's Healthy Bucks program and federal food assistance programs.

Intermediated Case: Hannibal School District

The Hannibal School District is located in the Syracuse MSA, near the largest apple-producing region in the state. It is in Oswego County, which has an annual per capita income of about $20,000, substantially lower than the average for New York. The school district has more than 1,600 students and includes three schools: elementary, middle, and high schools. Each has a cafeteria to serve the students. This school district has participated in various programs to increase consumption of fruit and vegetables among students and to link local farm products with school district cafeterias. According to the food service director, in recent years the school cafeterias have introduced a wide variety of healthy options on the menu, and apple purchases, as well as other produce items, have increased substantially. The share of apples sold through all school districts in the MSA is less than 1 percent of total MSA apple sales. It is difficult to identify the share of local apples in this channel.[6]

The food service director estimates that by the end of the 2009–2010 school year, the district will have offered about 15,000 pounds of apples to students. Most of these apples come from New York, except for a small proportion supplied to the school district by the Department of Defense (DOD) Fresh Fruit and Vegetable Program. These apples usually consist of the Red Delicious variety produced in Washington. Other than the DOD apples, the rest are produced on farms located in the same county as the school district or in adjacent counties. The school district has a warehouse where apples are received and subsequently distributed in school-owned vans to the three school cafeterias. Most apples offered in the school cafeterias are medium-sized, and the most common variety is the Empire apple, which was developed in New York.

Figure 5 traces the supply chain for apples served in the school district cafeterias. This supply chain consists primarily of four channel members who have maintained business relationships for over 20 years, the school district, a produce wholesaler (C's Farms), and two local farms. C's Farms supplies approximately 90 percent of all of the school district apples. The remaining 10 percent of the apples are those provided to the school for free by the DOD Fresh Fruit and Vegetable Program. The wholesaler, in turn, procures apples primarily from two apple farms located in the county, each with about a 50 percent share. As a price safeguard, the school district requests price quotes from a large national food service distributor as well as from its local supplier, C's Farms.

Production

Ontario Orchards is one of two apple suppliers from whom C's Farms buys. It is a family-owned and family-operated business dating back several

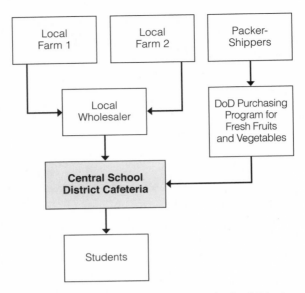

Fig. 5. Intermediated supply chain, New York Central School District.

generations, with 125 acres in production, including 90 acres in apple orchards. Annual apple production is about 300,000 pounds. The farm produces a large variety of fruits and vegetables, but apples are predominant, representing 80 percent of farm sales and including 29 apple varieties.

Ontario Orchards also operates a highly diversified farm stand that offers a wide selection of products and services: fresh produce, bakery, plants, a garden center, and landscaping. It employs nearly 70 people during the harvest season and about 30 during the rest of the year. Ontario Orchards apples are sold through the following channels: about 78 percent retail, 15 percent U-pick, 15 percent processed, and 2 percent through C's Farms. The apples sold to C's Farms represent about 40 percent of the school district's purchases.

The apples arrive from the orchards to the packing-shipping facility in bins. In the back of the farm market store, Ontario Orchards operates a production line in which apples are washed, sized, and packed in 42-pound boxes. The apples distributed to the school are U.S. No. 1 grade, medium size. The farm has a refrigerated room for short-term storage of apples in bins and in boxes. When asked about storage, the owner stated that the availability of long-term storage facilities in the area using the SmartFresh technology—an ethylene treatment for quality enhancement—has played a critical role in keeping the school district supplied with apples throughout the school year.

Wholesale Distribution

C's Farms delivers fresh fruits and vegetables to 65 local restaurants, schools, and institutions in Oswego County. It has been in business for more than 20 years and is family owned and operated. It employs 11 workers from September to March (when schools and colleges are in session), and six employees in other months. Although wholesaling is its primary activity, C's Farms also has a retail operation that offers fresh fruits and vegetables, fresh cut trees, ornamentals, and a wide variety of beers produced by micro-brewers. C's Farms is an active supporter of local farms, and during the summer and fall seasons most of the fruits and vegetables in its inventory are procured locally.

Apples account for about 7 percent of the wholesaler's sales, totaling approximately 200 bushels per month and yielding average monthly revenues of $3,700. About 18 percent of these apples (approximately 10 bushels per week) are sold to the Hannibal School District. Although its primary variety is the Empire, C's Farms sells other varieties such as Paula Red and Early Mac during the harvest season.

C's Farms participates in a variety of state programs aimed at strengthening local supply chains, including the Pride of New York Program and the Cornell Farm-to-School Program. It uses the Pride of New York logo on all its local products. This program supports market access to agricultural products grown and processed in the state and has played an important role in creating awareness for New York grown fruits and vegetables.

The School District

The school district serves 1,600 students at its three school cafeterias. The purchasing director is responsible for all purchases, including apples, and a school employee assists with the delivery of products from the warehouse to the school cafeterias.

About 35 percent of the school district's fresh produce purchases are apples, making them the largest produce item. Most apples offered in the school district are Empires because this local variety is strongly preferred by students, is part of the Farm-to-School Program, and holds well in storage. On some occasions, typically two or three times per year, the school district receives free apples from DOD Fresh Fruit and Vegetable Program. Although the apples are free, they are offered on an irregular schedule communicated to the food service director about two weeks in advance so that adjustments in local procurements can be made.

Apples are always one of the two produce items offered as part of a reimbursable meal at the school cafeterias. Students may purchase a lunch and then have the option to choose an apple as part of the meal or they may buy an apple separately, a la carte.

The apples sold in the school cafeterias are a smaller size and lower grade than is normally found in supermarkets. Apple prices and consumption exhibit very little variability throughout the school year, and the price paid to C's Farms is set for the harvest season based on market prices for New York published by USDA's Agricultural Marketing Service. The food service director and the wholesaler also meet in January to explore possible price adjustments for the spring, based on price projections of the same federal agency.

Recent years have seen an increase in apple consumption among students in the school district due to various promotional activities aimed at increasing the consumption of fresh fruit and vegetables among students, as explained in the following sections.

Marketing and Distribution

C's Farms receives orders from the school district's food service director on Thursday or Friday each week. It then consolidates the school district's order with the orders from other customers (it has 65 customers), and places a consolidated order with the two local farms, ordering about the same amount from each, on Monday or Tuesday the following week. In general, orders placed to Ontario Orchards on Monday or Tuesday are delivered to the wholesaler's warehouse on Friday or Saturday of the same week. It usually takes three to four days between placing the order with the grower and having it delivered to the warehouse, and about one week between receiving an order from the school and delivering it to the school. On some occasions, when local apple supplies are not adequate to meet the demand, the wholesaler orders apples from two other local farms or from additional suppliers either inside or outside New York.

No written contracts are employed between Ontario Orchards and C's Farms. According to the owner, a verbal contract has been in place for the past 20 years. Each order is invoiced separately, and every load is inspected at the warehouse upon delivery. Rejections of apples arriving at the warehouse are rare. When this happens the producer picks up the damaged product, replacing it practically immediately, without additional cost to C's Farms. Once an apple shipment is accepted, the invoice is signed and the grower receives payment from the wholesaler, normally in 30 days.

A primary objective of C's Farms is to ensure adequate supply to all its customers. The wholesaler maintains a small quantity of apples in cold storage, usually to meet a week's needs. Most apples, however, are stored by the supplier farms, which have access to short- and long-term storage with ethylene treatment, extending the life of the apples significantly. As a result, local apples are often available during the academic year.

Forecasts for apple requirements are the responsibility of the school's food service director as a function of the planned menu for the upcoming week. This process is relatively simple, since the consumption of fresh apples in school cafeterias is mostly constant over the school year. C's Farms keeps track of the weekly orders from its customers and projects its needs accordingly. On some occasions, when customers require more apples than usual (for example, for the New York Farm-to-School Day), they inform the wholesaler a month in advance, which gives the wholesaler enough time to adjust its orders.

For most products, the food service director follows a bid process with food service distributors. In the produce category, the director requests periodic quotes of prices and availability from a large national food service distributor and from C's Farms. With limited cold storage, the food service director generally orders apples from the wholesaler once a week. The local wholesaler is always the supplier for apples because of the price offered, the reliability of supply, the frequency of delivery, and the quality and freshness of local apples. The orders placed with C's Farms typically include between 15 and 16 produce items for the weekly menu.

Supply Chain Member Relationships

All supply chain members are located in the same county, and communications between them are informal and frequent. Word of mouth plays an important role in this supply chain's relationships, and C's Farms has a long history of business relationships with its suppliers. Formal contracts are nonexistent.

Ontario Orchards' owner values the wholesaler's knowledge of standards, frequent communication, and timely payments. When asked, the C's Farms owner stated the most valued aspects of its relationship with suppliers are continuous open communication, the level of service provided, and the freshness of the products delivered from local farms. The owner also emphasized that supporting local farmers and businesses is key to maintaining good relationships among supply chain members.

The food service director values C's Farms' consistent, reliable supply and responsiveness to quality issues (e.g., replacing product immediately if quality problems are encountered). C's Farms frequently shares apple market reports from USDA's Agricultural Marketing Service with the food service director, contributing to transparency in setting prices. C's Farms, for its part, values fast payment from the school district and the good word-of-mouth messages from the food service director to other local businesses.

Prices along the Supply Chain

We calculate the retail price for apples as the wholesale price times 2.25—the markup rule employed by the school district. In 2009 the average price for medium apples paid by the school district to C's Farms was $0.42/lb, so the retail price was $0.90/lb ($0.42 x 2.25 = $0.90) and the school district's share of retail revenue was $0.48/lb (53 percent). There are no documented transportation costs along this supply chain given the proximity of chain members. The average price for medium apples paid by C's Farms to the two suppliers at the farm gate was $0.32/lb, so the shares of retail revenue were $0.10/lb (11 percent) and $0.32/lb (36 percent) for C's Farms and Ontario Orchards, respectively.

Food Miles and Transportation Fuel Use

Ontario Orchards is located within 10 miles of C's Farms facilities, and the average weekly apple order from C's is about 30 bushels. These batches are either picked up by the wholesaler or delivered to the warehouse by the farms. Transportation costs are not specifically charged, given the proximity to the wholesaler's facilities. The order usually consists of apples only and is delivered in a box-van. Both suppliers deliver apples that are washed and sized in cardboard boxes to the wholesaler.

Apples are delivered at no cost by C's Farms to the receiving facilities at the school district. They are transported with other produce items in a box-van. The apples and all other produce items are subsequently distributed among the three cafeterias by school district personnel.

Community and Economic Linkages

The food service director is enthusiastic about activities to promote apples and to support local farms. The school district has had programs in place for several years. For example, each Wednesday is called the "Fresh Food Day" and apples are offered in a large bowl in the cafeterias while the servers encourage students to consume them. For several years the school district has participated in promotions such as the program "An Apple a Day" (AAD). AAD was launched by the Addictions Care Center of Albany, a private nonprofit organization that helps maintain critical programs for the prevention of alcohol and substance abuse, emphasizing healthy habits in children. The school district has also participated in "Fruit for Dessert" programs with an emphasis on apples.

In recent years school activities to promote fruit and vegetable consumption have increased dramatically. In 2009 the school nutrition team launched a program called the "Smart Choice Café!" in collaboration with four other school districts in the county. The food service directors collaborated with a registered dietitian to design the school menu, to include more nutritious, tasty, and affordable items, and to plan promotional campaigns among students to increase fresh fruit and vegetable consumption. The team also employs a menu flier to promote produce and make produce consumption "cool." This flier includes pictures of apples and other fresh fruit and vegetables, highlighting their health benefits and nutritional properties. Moreover, all servers in the cafeterias are instructed to remind each student "not to forget your fruit and vegetables."

School Support for Locally Grown Fruits and Vegetables

The food service director collaborates with the county's Cornell Cooperative Extension office and with the Oswego County Farm Bureau for promotion of local fruits and vegetables. She also participates in a variety of events aimed at promoting local produce (and apples), including organizing Health Fairs, maintaining school gardens, and hosting educational dinners featuring local foods. The following are two of the most important programs:

> The New York Harvest for New York Kids Fest takes place in early October each year. During a complete week, the menus focus on New York-produced foods and, on one particular day, the menu consists of only New York foods. This event allows schools and communities to learn about New York agriculture, develop preferences for locally-grown foods, and make healthy food choices. During this event, classrooms do food-tastings and students visit local farms and harvest their school gardens.
>
> Cornell's Farm-to-School Research and Extension Program was established in 2002 to initiate, develop, and sustain farm to school connections. This program supports efforts to increase the amount of locally-produced fresh or minimally-processed foods served primarily in New York's educational institutions. The Program conducts research, provides educational opportunities, and offers technical assistance to extension educators, nutrition professionals, food service directors, farmers, parents, and students interested in developing farm to school connections. The program has a website serving as an informational hub for information about farm to school interests, opportunities, challenges and activities across the state (Cornell University Farm to School Extension and Research Program 2011).

Apples from the DOD Fresh Fruit and Vegetable Program

The DOD Fresh Fruit and Vegetable Program started in 1995 and is based on the principle that the Department of Defense can deliver fruit and vegetables directly to schools along with deliveries to military installations. The program was well received by many states and grew from a pilot program delivering $3 million of produce to eight states in 1995 to distributing $20 million of produce nationwide in 1999. The 2002 Farm Bill more than doubled this spending. States or their schools place orders directly with DOD field offices for a wide variety of American-grown specialty crop products. The program is so important for specialty crops that by 2004 USDA had become the second largest customer of the Department of Defense.

In the case of the focal school district, the food service director does not make decisions regarding the sourcing of DOD apples. The decision is made at the state level. The New York Department of Agriculture and Markets orders the DOD Fresh Fruit and Vegetables Program apples for the state. The food service director orders DOD Fresh Fruit and Vegetables Program apples two or three times each academic year, but there is not a pre-established schedule for deliveries, since orders depend on product availability. These apples arrive in brown boxes and are free for the school district. Of course, free apples are beneficial for the school budget, particularly in tight budgetary times. All channel members interviewed stated that this program poses coordination challenges for the local apple supply chain because DOD Fresh Fruit and Vegetables Program apples represent about 10 percent of the apples consumed in the school cafeterias.

Prospects for Expansion of Local Apples in School Districts

This case illustrates several factors that support the expansion of apple supply chains for school districts. Perhaps the most important is the community support for local farmers, especially in the case of C's Farms and the school district. In addition, apple sales can continue to benefit from increased diet-related health concerns that have prompted the inclusion and promotion of fresh, healthy items in school menus. Third, the existence of a cold storage infrastructure and the technology to have local apples available throughout the school year has facilitated the expansion of apples in school menus. Finally, collaborative arrangements among distribution channel members ensure a reliable supply of apples. The future expansion of local apples in other school districts depends on the ability to link educational efforts to improve the nutritional intake of pupils with the availability of reliable supply chains for local apples.

At the same time, this case identifies several factors hindering expansion of apples in the school district. The population in the Syracuse MSA, including children of school age, is declining. Therefore the school district market for apples may not be of sufficient volume to support the expansion of this particular local supply chain. Second, apples distributed through the school district represent a very small portion of the total apple market in the metro area. For example, while the focal school district uses about 15,000 pounds of apples per year, annual apple sales for the focal supermarket in Syracuse exceed 1.5 million pounds. Finally, tighter budgets may be an impediment to including healthier options in the school menu. Fresh fruits and vegetables are expensive in comparison to other food items.

Key Lessons

This case study provides some unique lessons regarding the apple supply chain in school districts and the role of this distribution channel in the overall apple production and distribution system.

First, the geographic location of this school district facilitates the existence of a sustainable local supply chain. New York, and in particular the school district, is in an apple-producing region with a large number of small and medium-sized apple growers who utilize the school district to diversify their distribution strategy. In addition, the state has the required post-harvest infrastructure to store apples for year-round availability.

Second, the role of the wholesaler is critical. C's Farms assembles the product, receives and consolidates orders, and distributes to various buyers. These supply chain functions are not being conducted by the local growers or the school district. This finding highlights the importance of an intermediary to aggregate the products, particularly in supply chains that involve a large number of small producers. The viability of this supply chain would be in question without the functions performed by C's Farms.

Third, the market for apples in the school district is small relative to the market in other distribution channels (e.g., supermarkets). This case suggests that chances are slim that school districts will become the primary distribution channel for apple growers. Most likely, school districts will remain a distribution channel for apples that do not have the size or quality required for retailers' shelves, thus providing an appropriate outlet for apples of this type.

Fourth, government support through such initiatives as the Farm-to-School Program seems critical for the supply of local apples to the school district. Nevertheless, the case shows a lack of coordination between at least two federal programs, the Farm-to-School Program and the DOD's Fresh Fruit and Vegetable Program. This case shows that the DOD Fresh Fruit and Vegetable Program may be hindering the development of this local supply chain by offering free, nonlocal apples to the school district on a sporadic basis.

Cross-Case Comparisons

The three case studies conducted in the Syracuse MSA—SuperFoods, Jim Smith Farm, and the Hannibal School District—illustrate several of the many supply chains for fresh apples. This final section compares the three supply chains for apples in Syracuse in terms of their differences and similarities regarding supply chain structure, size, and performance and concludes with a brief discussion of key lessons emerging from the three cases.

Supply Chain Structure

Our cases identify many similarities and differences across direct, intermediated, and mainstream supply chains. We examine six specific aspects about the structure of local food supply chains using the mainstream case as a point of comparison.

Only Jim Smith Farm, through participation in the direct market supply chain, provides the consumer with information about where and by whom apples were produced. The mainstream chain provides information about the state where apples were produced for some but not all apples. The school district conveys information about where and by whom the apples are produced but only indirectly, via community events such as the New York Harvest for New York Kids Festival.

Durable relationships, characterized by a high degree of trust, information sharing, and decision sharing over time, exist between supply chain partners in the mainstream and intermediated chains. This indicates that trading partners exhibit strong mutual interdependence in the mainstream and intermediated apple supply chains. SuperFoods (the focal firm in the mainstream supply chain) and the Hannibal School District (the focal organization in the intermediated supply chain) have had business relationships based on longtime trust and frequent communication with their respective business partners without requiring formal contracts. We also found that information sharing is extensive and frequent in the mainstream supply chain. Apple suppliers and SuperFoods share supply chain information electronically nearly daily, including inventory and product availability. Although information is also shared on a limited basis between vendor and customers in the farmers market case, the relationships are not as durable as in the other two supply chains.

Prices received by producers in the SuperFoods and the school district supply chains are primarily determined in commodity markets. In the SuperFoods case, however, producers are vertically integrated grower-packer-shippers and receive prices that reflect these value-added activities. Although apple prices received by Jim Smith at the farmers market are not determined in commodity markets, he faces a high degree of competition in the direct supply chain and therefore behaves as a price taker.

Collective organizations play an important role in the direct market supply chain. The CNY Regional Farmers Market and the other farmers markets, all collective organizations, have facilitated the participation of Jim Smith in direct market supply chains. In addition, New York hosts active trade associations that promote direct marketers such as the Farmers' Market Federation of New York. In contrast, collective organizations do not play a significant role in the SuperFoods and school district apple supply chains.

The cases demonstrate that the presence of a strong industry distributing nationally or internationally helps to create an infrastructure of knowledge and services that facilitates the development of direct and intermediated food supply chains. New York ranks second in apple production nationally, and the industry distributes nationally and internationally. This has contributed substantially to the development of direct and intermediated supply chains for local apples. For example, lack of year-round availability is not an issue for apples in direct and intermediated supply chains due to large statewide production volume and due to storage technology allowing year-round distribution. Moreover, producers in direct and intermediated chains are able to diversify their distribution channel strategy and can also sell their harvest through mainstream channels (e.g., New York packer-shippers).

Jim Smith has benefited from the CNY Regional Farmers Market and from other thriving farmers markets in the region. Ontario Farms runs a very successful retail operation with a wide variety of products procured locally, other than apples. Finally, SuperFoods has developed a locally grown program with participation of over a thousand farmers. Through this program SuperFoods delivers a variety of educational programs for local producers, including GAP and food safety training. Additionally, the school district has been active in helping to foster the distribution of local foods in the community.

Supply Chain Size and Growth

Case study findings provide information about critical issues influencing supply chain size and growth.

The mainstream supermarket supply chain has by far the largest market share in Syracuse. The focal store sells 1.5 million pounds of apples per year. The combined sales volume of direct and intermediated chains represents a small proportion of total apple demand. Jim Smith sells approximately 40,000 pounds of apples per year in the farmers market, a volume similar to that of the other 8–10 apple vendors in the direct supply chain, and we estimate that less than 1 percent of all apples in the Syracuse MSA are sold in this direct chain. Likewise, the volume of apples sold in school districts is small, and the share of apples sold through school districts in the MSA is less than 1 percent of total MSA sales. In the case of local apples procured by school districts, the Hannibal School District procures approximately 15,000 pounds of apples during the academic year. Not all the school districts in the Syracuse MSA have programs to buy apples from local producers.

Access to and costs associated with processing and distribution services are not currently a constraint on sales volume in any of the three case study supply chains. The direct and intermediated supply chains target low-volume distribution channels (a farmers market and a school district) that do not compete directly with the mainstream supply chains.

All apple suppliers in the mainstream supply chain have high quality and service standards to ensure the product consistency, quality, and volume required by SuperFoods. Costs of compliance with quality and safety standards are not an issue for local apples in either school cafeterias or SuperFoods. We note, however, that the DOD Fresh Fruit and Vegetable Program may limit the development of local supply chains supplying apples to the school district.

While public regulations are not currently limiting the ability of smaller producers to enter mainstream channels, retailers are starting to require third-party food safety certification of all suppliers. While none of our case study subjects' apples have ever been implicated in any food safety recalls, food industry members and food safety experts speculate about certain future blanket requirements for food safety certification for all growers selling in mainstream channels.

Lack of year-round availability is not an issue for local apples in any of the chains. This is due to New York's strong presence in the national market, large statewide production volume, and availability of storage technology. Summer is a challenge for all chains because the quality of imported apples, depending on the variety, can be superior to certain domestic apples kept in long-term storage.

Apple sales in the direct market chain are expanding primarily through entry of new firms, as more vendors sell apples in the farmers market during the apple season. However, Jim Smith has been able to expand his direct distribution by participating in other farmers markets operating in the region. Ontario Orchards has responded to increased demand for local apples through internal growth, by diversifying its distribution channels, including U-Pick, retail, and wholesale distribution.

Supply Chain Performance

Finally, we examine how direct market and intermediated supply chain performance compares with the performance of mainstream chains.

Table 3 shows how the allocation of revenues to producers varies across supply chains. Producer share of the price paid by the final consumer is greatest for the direct market chain, as expected. The price received by

Jim Smith, net of marketing expenses, is $0.40/lb or 80 percent of his direct market retail price. Marketing expenses for the direct marketing chain are estimated to total $0.10/lb or 20 percent of the retail value. We use the USDA average grower price, $0.26/lb, as a proxy for the grower price for our integrated grower-shipper-packers. Using this figure, the grower-packer-shipper share of the price paid by the final consumer is lowest for the intermediated supply chain (Ontario Orchards, 36 percent) and for the mainstream supply chain's Washington suppliers (GPS3, 37 percent). Since these businesses also incur packing and shipping costs, marketing expenses are added to producer price for these businesses. Finally, New York GPS1's share of the final price paid by the consumer is 47 percent for bulk apples and 60 percent for bagged apples.[7]

A critical issue for all these supply chains is whether differentiation by quality attributes other than "local" is necessary to receive and sustain price premiums for local food products. Comparing apple prices is not easy due to quality and size differences as well as the large number of commercial varieties that sometimes enjoy different price points. Therefore, examining the price premium of the "local" attribute is not straightforward. In figure 6 we present weekly prices from our case study participants for selected varieties, packaging, and case study outlet. The main variety supplied by Washington State supplier GPS3 is Red Delicious, sold in bulk in SuperFoods. The main variety supplied by New York supplier GPS1 is Empire, sold both in bulk (of a larger size and higher grade) and in bags (of a smaller size and lower grade). And the main variety sold in the school district is also Empire, sold only in bulk (of the smallest size and lowest retail grade). All apples sold by Jim Smith in the farmers markets are sold in tote bags, and they are not graded or sized but are sorted by hand to remove those the farm does not want to sell through the market.

The retail price for bulk apples is lowest in school district cafeterias because of the lower grade and size. Interestingly, the local bulk apples (Empire) supplied by GPS1 to SuperFoods had lower prices than bulk apples (Red Delicious) grown in Washington and supplied by GPS3. Although different in variety and grade, this higher price for Washington apples reflects in part the pricing strategy of SuperFoods to promote New York apples. It is of note that the CNY Regional Farmers Market offers the

Table 3. Allocation of retail revenue in Syracuse, New York—apple chains, by supply chain and segment

Supply chain segment	MAINSTREAM[a] SuperFoods GPS3 (WA-Bulk) Revenue ($/lb)	% of total	MAINSTREAM SuperFoods GPS1 (NY-Bulk) Revenue ($/lb)	% of total
Producer[c]	0.26	14	0.26	17
Producer estimated marketing costs[d]	na	na	na	na
Packer-Shipper	0.40	21	0.45	30
Transport	0.23	12	0.03	2
Wholesaler	na	na	na	na
Retailer	1.00	53	0.76	51
Total retail value	1.89	100	1.50	100

[a] GPS1 and GPS3 are grower-packer-shippers; SuperFoods is a wholesaler-retailer. Average prices using SuperFoods records from August 2008 through July 2009 for the direct and intermediated cases.

[b] The producer, Ontario Orchards, is a grower-packer-shipper in the school district supply chain.

[c] Producer prices are the monthly average for the period 2000–2008 reported by the U.S. Department of Agriculture, usda.mannlib.cornell.edu/MannUsda/viewDocumentInfo.do?documentID=1377 (accessed 12/14/2011).

[d] Includes estimated costs of farmers market stall fees, transport to market, the opportunity cost of family labor, and tote bags for customers. Total producer per unit revenue is 0.40 + 0.10 = 0.50 ($/lb).

[e] 95 percent of apples in the school district are sold as part of the school menu and thus do not have a specific retail price. We calculate the retail price as the wholesale price times 2.25 ($0.42 x 2.25 = $0.90), the markup rule employed by the school district.

Source: King et al. (2010: 15).

MAINSTREAM		DIRECT		INTERMEDIATED[b]	
SuperFoods GPS1 (NY-Bagged)		Jim Smith Farm		School District	
Revenue ($/lb)	% of total	Revenue ($/lb)	% of total	Revenue ($/lb)	% of total
0.26	26	0.40	80	0.26	29
na	na	0.10	20	na	na
0.34	34	na	na	0.06	7
0.03	3	na	na	na	na
na	na	na	na	0.10	11
0.37	37	na	na	0.48	53
1.00	100	0.50	100	0.90[e]	100

lowest-priced apples among all supply chains and about half the price of comparable apples in the mainstream supply chain. These factors suggest that attributes other than "local" (e.g., variety, size, and grade) are necessary to receive and sustain price premiums. The attribute "local" does not appear to command a price premium in the case of apples.

The apple cases suggest that direct and intermediated supply chains contribute a greater portion of their wage and business proprietor income to local economies than do mainstream supply chains. All members in the school district supply chain are local, and practically all wages and proprietor income stay within the school district zone and its community. In the mainstream supply chain roughly one-third of the wage and business proprietor income is realized outside the region, because Washington supplies about one-third of the chain's apples. However, the remaining two-thirds of the wages and income of this supply chain produce a large volume of wages and proprietary income to the local economies simply due to the size of the business entities.

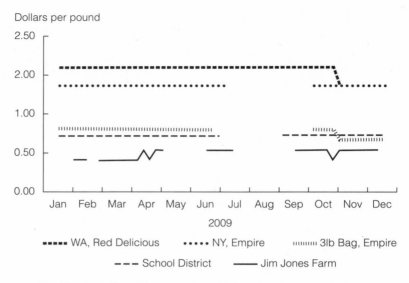

Fig. 6. Weekly prices for apples in various retail outlets, 2009. Source: Authors' interviews and company records.

Product miles traveled and fuel use for transportation per unit of product sold is another critical indicator of supply chain performance. The cases indicate that apples travel the fewest miles in the intermediated supply chain, followed by the direct market supply chain (table 4). Apples supplied from Washington state travel the most miles. However, fuel use per 100 pounds of product illustrates other differences across supply chains. The intermediated supply chain has the highest fuel efficiency. The direct supply chain and the supply chain formed by local G P S 1-Superfoods have comparable fuel use per 100 pounds of product (0.16 and 0.20 gallons/cwt, respectively). Apples supplied by the mainstream G P S 3 in Washington have the worst fuel usage performance (1.41 gallons/cwt). These findings suggest that intermediated and direct supply chains can be fuel efficient.

Regarding the creation of social capital and civic engagement in the consumption area, all three supply chains engage in a variety of activities to establish communities of producers and/or consumers. The mainstream chain engages in several events to support communities near the focal store, provides extensive training to local growers, and promotes visits of employees in the produce department to supplier farms. The direct market and local intermediated chain (i.e., apples sold to the school district) also engage in various community-building activities, including the promotion of local apples in school districts and farmers markets. The focal firms in all three supply chains participate in civic organizations, but the types of organizations differ considerably across the chains. The mainstream chain tends to participate in national organizations, whereas the intermediated and direct market chains tend to participate in local or regional associations. Finally, our mainstream and direct market cases indicate that direct marketing helps build entrepreneurial skills. Specifically, SuperFoods has provided extensive training to local growers to participate in their local produce program, strengthening the entrepreneurial skills of specialty crop growers in general and apple growers in particular.

Key Lessons

Four general lessons emerge from these case studies of apple supply chains in Syracuse, New York:

1. The three case studies indicate that members of the apple supply chains exhibit a high degree of diversification in their distribution channels. Local and mainstream apples coexist and complement one another in the supermarket channel. The focal farmers market vendor engages in some direct marketing but is also linked to the mainstream chain through his relationship with a conventional packer-shipper. Meanwhile, the school district procures both from food service suppliers and from local apple supply chains. Local supply chains are profitable and important for participating firms even if the volume is small.
2. The presence of a strong industry that distributes nationally has substantially facilitated the development of local food supply chains. The New York apple sector offers a wide variety of products to consumers regionally and nationally, and as a result it has the post-harvest infrastructure (e.g., packing, shipping, short- and long-term storage) and marketing expertise to support distribution of apples from local farms to various local retail and food service outlets.
3. The case studies underscore the high degree of competition within the apple sector as reflected by the price formation mechanisms. Final prices are generally established by the market in all supply

Table 4. Food miles and transportation fuel use in Syracuse, New York—apple supply chains

SUPPLY CHAIN SEGMENT	FOOD MILES	TRUCK MILES	RETAIL WEIGHT (CWT)	FUEL USE (GAL)[a]	FUEL USE PER CWT SHIPPED
Mainstream: SuperFoods, GPS3 (Washington)					
Producer to Packer-Shipper	150	300	100	25.0	0.25
Packer-Shipper to Distribution	2,600	2,600	400	433.3	1.08
Distribution to Retail[b]	100	200	400	33.3	0.08
All Segments	2,850				1.41
Mainstream: SuperFoods, GPS1 (New York)					
Producer to Packer-Shipper	25	50	100	4.2	0.04
Packer-Shipper to Distribution	100	200	400	33.3	0.08
Distribution to Retail	100	200	400	33.3	0.08
All Segments	225				0.20
Direct: Jim Smith Farm[c]					
Producer to Retail	61	122	20	3.1	0.16
All Segments	61				0.16
Intermediated: School District[d]					
Producer to Wholesaler	3	6	10.0	0.3	0.03
Wholesaler to School District	10	20	40.0	0.3	0.01
All Segments	13				0.04

[a.] Miles per gallon vary by segment. Trailer trucks shipping apples from packing shed to the distribution center have capacity of 40,000 lb and obtain 6 mpg; trucks used to transport apples from the farm to the packing shed have capacity of 10,000 lb and obtain 12 mpg.

[b.] Apples are about 5 percent or 20 cwt of the total weight of products transported in trailer trucks from the distribution center to the store. These trucks have capacity of 40,000 lb and obtain 6 mpg.

[c.] The box-van employed in the direct market supply chain has capacity of 2,000 lb and obtains 20 mpg.

[d.] The box-van employed from the producer to the wholesaler transports 1,000 lb and obtains 20 mpg; the truck employed to transport apples from the wholesaler to the school district has capacity of 4,000 lb and obtains 20 mpg.

Source: King et al. (2010: 16).

chains considered, with the exception of a few truly uncommon apple varieties in the farmers market produced in very small quantities. In all supply chains, apple growers appear to be price takers. It is noteworthy that we observed no price premiums for local apples in any of the supply chains studied. We speculate that because New York is a major apple producer with year-round supplies, "local" is not a significant differentiating attribute.

4. Our case studies highlight that growers tend to retain a greater share of the price paid by the final consumer in the direct market chain but not in the intermediated supply chain. The local and nonlocal supply chains coexisting in SuperFoods suggest that local grower-packer-shippers retain a greater share of the price paid by the end consumer than do their Washington state counterparts. Finally, when comparing the mainstream and direct market cases, we find that the longer the distance and the more intermediaries between the farm and the point of sale, the smaller the share of the final consumer price that is retained by the grower.

Notes

1. See agriculture.ny.gov/AP/CommunityFarmersMarkets.asp.
2. This is a pseudonym as the firm prefers to have its identity remain confidential.
3. Apples travel from less than one mile to as far as three hundred miles from the farm gate to the packing house.

4. This is a pseudonym as the firm prefers to have its identity remain confidential.
5. Average of the monthly producer prices for the period 2000–2008 reported by the U.S. Department of Agriculture (USDA-ERS n.d.b).
6. Annual apple consumption by students in all school districts in the MSA was estimated to be 847,987 pounds, by extrapolating the consumption of 1,600 students in the Hannibal School District (15,000 pounds) to the population of the MSA aged between 5 and 18 years, 90,452 (U.S. Census Bureau n.d.): that is, (15,000/1,600) × 90,452) = 847,987.
7. These suppliers are integrated grower-packer-shippers. The share of the retail dollar for Washington supplier GPS3 is the summation of several supply chain segments in table 1, namely 14 + 21 = 35 percent; for New York supplier GPS1-bulk it is 17 + 30 = 47 percent; and for New York supplier GPS1-bagged it is 26 + 34 = 60 percent.

References

Central New York Regional Market Authority. 2010. *2010 CNY Regional Market Authority Vendor Handbook*. cnyrma.com/pdf/vendorbook_web.pdf (accessed 12/8/2011).

Cornell University Farm to School Extension and Research Program. 2011. "Cornell Farm to School Program." farmtoschool.cce.cornell.edu/ (accessed 10/30/2011).

Farmers' Market Federation of New York (FMFNY). 2006. "The Value of Farmers' Markets to New York's Communities." nyfarmersmarket.com/pdf_files/Farmers MarketsCommunityDev.pdf (accessed 12/8/2011).

King, Robert P., Michael S. Hand, Gigi DiGiacomo, Kate Clancy, Miguel I. Gómez, Shermain D. Hardesty, Larry Lev, and Edward W. McLaughlin. 2010. *Comparing the Structure, Size, and Performance of Local and Mainstream Food Supply Chains*. ERR-99. Washington DC: U.S. Department of Agriculture, Economic Research Service.

Merwin, Ian A. 2008. "Some Antique Apples for Modern Orchards." *New York Fruit Quarterly* 16(4): 11–17.

U.S. Census Bureau. N.d. "Population and Housing Unit Estimates." census.gov/popest /index.html (accessed 12/8/2011).

USDA-ERS. N.d.a. "Organic Production Data Sets, State-level Table, Table 11." ers.usda .gov/data/organic (accessed 12/8/2011).

———. N.d.b. "Fruit and Tree Nut Yearbook Spreadsheet Files." usda.mannlib.cornell .edu/MannUsda/viewDocumentInfo.do?documentID=1377 (accessed 12/8/2011).

USDA-NASS. 2007. "Agricultural Census: United States Summary and State Data. Volume 1. Chapter 2." U.S. Department of Agriculture, National Agricultural Statistics Service. agcensus.usda.gov/Publications/2007/Full_Report/Volume_1,_Chapter_2 _us_State_Level/st99_2_002_002.pdf (accessed 12/8/2011).

———. 2009. "Noncitrus Fruits and Nuts 2009 Preliminary Summary." U.S. Department of Agriculture, National Agricultural Statistics Service. usda.mannlib.cornell .edu/usda/nass/NoncFruiNu//2010s/2010/NoncFruiNu-01–22–2010_revision.pdf (accessed 12/8/2011).

4   Blueberry Case Studies in Portland-Vancouver MSA

Larry Lev

Introduction

This set of case studies describes three fresh blueberry supply chains in the Portland-Vancouver MSA (referred to as Portland):

- a major supermarket chain supplied in part by a local grower-packer-shipper (mainstream supply chain),
- a producer who sells through farmers markets and farm stands (direct market supply chain), and
- a regional natural foods store chain that features locally produced berries (intermediated supply chain).

Products that are produced, processed, and distributed in Oregon and Washington (referred to as the Northwest) are defined as "local." Supply chains that distribute local products and convey information that enables consumers to identify the products as local are considered "local food supply chains."

The Location: Portland-Vancouver MSA

The Portland metropolitan area (figure 7) straddles two states and has a population of 2.2 million. It consists of the counties of Multnomah, Washington, Clackamas, Columbia, and Yamhill in Oregon and Clark and Skamania Counties in Washington. The Oregon counties in the MSA represent 47 percent of the state's population, while the two Washington counties represent 6 percent of that state's population. In 2009, per capita

Portland metropolitan statistical area

Fig. 7. Portland MSA.

income of $38,936 for the MSA was almost identical to the national average of $39,357 (Oregon Employment Department 2011).

In 2005 metro Portland ranked as the 20th largest U.S. grocery market, with food sales of more than $4.3 billion. The Portland marketplace features the top five national food retailers, with Safeway and Fred Meyer (Kroger) having the largest share of food sales (Beaman and Johnson 2006). In addition, there are numerous regional food retailers and food cooperatives in the area.

A total of 11,692 or 18 percent of Northwest farms sold nearly $100 million through farm-direct supply chains in 2007, with average per farm sales of $8,552. By comparison, 6 percent of U.S. farms sell farm-direct, with average sales of $8,904 per farm. Northwest consumers purchased just over $10 per capita through farm-direct supply chains in 2007, or 2.5 times the national average of $4 for that year (USDA-NASS 2009b).

The Portland area supports 40 farmers markets in addition to many farm stands and consumer supported agriculture farms. Most markets are seasonal rather than year-round, and sales are highest during the summer months. Strong direct to restaurant, retail, and institution sales are supported by numerous private and public sector initiatives.

The Product: Fresh Blueberries

Consumer demand for blueberries has increased significantly over the past three decades due to favorable publicity related to their health benefits. The value of U.S. farm cash receipts for blueberries grew more quickly during 1980–2008 than for any other fruit—twelvefold as compared to threefold for the overall category (USDA-ERS 2009). North American domesticated blueberry production increased more than fivefold from 70 million pounds in 1968 to more than 400 million pounds in 2008 (U.S. Government Printing Office 2009: 36956). Globally the growth has been even faster, and the U.S. Highbush Blueberry Commission (n.d.) estimates that world production will more than double between 2008 and 2015. U.S. imports increased rapidly, and fresh blueberries are now available in supermarkets nearly year-round as new domestic and foreign sources fill the gaps in the traditional production calendar. The U.S. fresh market production season begins in California and southern states in April and extends into September. Argentina exports blueberries to the U.S. market in October and November, while Chile is the primary supplier for December through March. Although annual per capita fresh blueberry consumption is still only 1 lb/person, this represents a tripling in the past decade (Pollack and Perez 2009).

As a summer season crop, blueberries are a popular signature item in Northwest farmers markets and farm stands.[1] Availability of the local crop is limited to 10 to 12 weeks (July through September), as fresh blueberries cannot be stored for long periods. Oregon and Washington ranked third and fifth, respectively, in 2009 for cultivated blueberry production, and the Northwest provides just fewer than 24 percent of domestic production.

All the blueberry supply chains featured in these cases are located in Oregon, so some background information on Oregon agricultural production and the relative stature of the blueberry industry is useful.[2]

Oregon has a diversified agricultural sector with total farm sales of $4.5 billion and more than 100 crops with sales in excess of $1 million. Berry crops with sales of $146 million in 2008 represent 3 percent of all farm sales. Blueberry sales of $49 million are the largest of the berry crops (USDA-NASS 2009a). The Oregon blueberry industry consists of 300 growers and 30 distributors who are registered with the Oregon Blueberry Commission (oregonblueberry.com/). Most producers are in the northern Willamette Valley in western Oregon, but there are some blueberry acres farther south in the state. Both Oregon State University and the U.S. Department of Agriculture have active research programs focused on production issues including varietal development.

Because the Northwest industry produces far more than its residents purchase, the region supplies both fresh and processed blueberries to consumers elsewhere in the United States and the world. Organic production represents the primary means for differentiating blueberries. Certified organic blueberry acreage, which has increased rapidly in the Northwest, now represents 9 percent of total blueberry acreage in the region (Kirby and Granatstein 2009a, 2009b).

Pack size has an important influence on retail prices for conventional blueberries sold in Portland. The median annual price of blueberries sold in packs of less than a pound is $8.00/lb, while the median annual price of blueberries sold in packs of greater than one pound is $5.32, or 33 percent less. This lower price reflects two factors at work. First, the larger pack sizes are primarily available when prices are already low. Second, consumers expect and generally receive a lower price for making larger purchases.[3]

Mainstream Case: Allfoods

Allfoods, a national supermarket with more than 1,000 stores including many in Portland, is representative of the primary way that most Portland area consumers purchase fresh blueberries.[4] The Allfoods produce department sells fresh blueberries throughout most of the year using domestic and international sources that change by season. Over the course of 2009, Portland Allfoods stores sold berries with 20 different brand names and often sold multiple brand names simultaneously.[5]

Hurst's Berry Farms with headquarters in Sheridan, Oregon (52 miles south of Portland) is a privately held business that produces, packs, and ships a wide variety of berries for both domestic and foreign customers (hursts-berry.com/). The blueberry component of the business includes the firm's own acreage in Oregon and Mexico, packing facilities in Oregon and California, and distribution agreements with Chilean and Argentine producers. As a result, Hurst's distributes blueberries virtually year-round.

Hurst's founder, Mark Hurst, participates in all the major blueberry industry groups at the state and national level, frequently in leadership roles. He has witnessed and participated in the rapid growth of the fresh blueberry industry.

Supply Chain Structure and Size

Product flows within the supply chain that provides Allfoods with local season production are shown in figure 8. Allfoods purchases blueberries from many different packer-shippers over the course of the year and during the Northwest production season. Hurst's represents a key supplier, especially during the Oregon production season. As shown in the figure, Hurst's packs and ships berries from its own production and also from other Oregon producers. The berries pass through the Allfoods distribution center before they are transported to the individual stores.

Production

Established in 1980 when Mark Hurst purchased a small farm in Oregon, Hurst's grows blueberries on 75 acres in Oregon.[6] Blueberry production and distribution represent about 60 percent of Hurst's revenues, with the rest of the revenues coming from the production and distribution of blackberries, raspberries, kiwi berries, currants, and gooseberries. The firm's diverse offerings and ability to provide products throughout the year are key characteristics. Hurst's does not produce or distribute any organic berries.

Hurst's is one of the bigger Oregon grower-packer-shippers, handling about 10 percent of all the blueberries produced in Oregon and about 25 percent of the fresh market blueberries. While a big player on the West Coast, Hurst's is much smaller than the largest national blueberry packer-shippers.

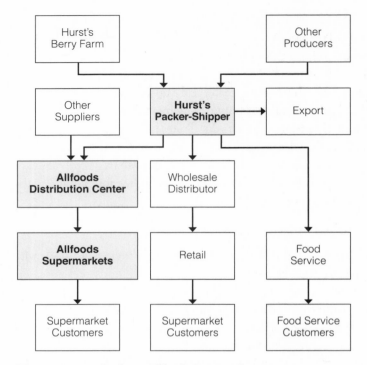

Fig. 8. Mainstream supply chain, Allfoods Supermarket.

The farm's own acreage provides between 5 and 10 percent of the Oregon blueberries marketed by Hurst's, with the rest being purchased from approximately 40 Oregon growers. Overall blueberry yields for the production that Hurst's handles are similar to the statewide average of 8,000 lbs/acre for fresh market blueberries (USDA-NASS 2008). Previously, Hurst's purchased from an even greater number of Oregon growers, but the company recognized that it could gain efficiency by reducing the number of lots going through the packing facility, since each lot is kept separate for control purposes.

Although machine harvesting of blueberries is far less expensive and is used to harvest almost all berries for processing, quality concerns continue to rule out machine harvesting for fresh berries. As a result, the main production challenge that fresh market growers face is having sufficient labor at harvest time. Hurst's addresses this challenge head on by operating a separate farm labor contracting business that employs about 100 farm workers and moves them around from the corporate farm to supplier farms as needed.

In most instances there are no formal contracts, but the 40 Oregon farms that supply Hurst's during the Northwest production season do so consistently, and both sides value the relationships formed over the years. This means that neither the farms nor Hurst's approach a new year with the idea of changing their existing partnerships. Individual farms do occasionally decide to make a permanent shift from one packer to another, so it is in this context that competition among Northwest packers needs to be understood.

Packing and Distribution

Oregon-produced blueberries that enter the mainstream distribution chain are handled much differently from those destined for direct or intermediated markets. Berries arrive in bulk at the packing facility from either the home farm or one of the supplier farms and are cooled down and packed within 24 hours. During the packing process the berries pass through three sorting lines, where a combination of mechanical and human assessment separates out substandard (soft or off-color) berries, and the remaining berries are packed into one of the six standard clam shell sizes that Hurst's offers (special requests can also be met). This whole process is much more thorough and automated than the process for berries that move through the direct and intermediated chains. In most instances it is also somewhat slower in transporting berries from the field to the final consumer. Once the Oregon season is finished, the sophisticated packing lines are dismantled and shipped to California so that they can be used for a greater portion of the year. The blueberries that Hurst's imports are packed before they are shipped, so they do not go through the Hurst's packing houses.

Recently Hurst's invested in controlled atmosphere storage and has instituted a system of selecting the highest quality fruit to be placed in this storage. The fruit placed in storage receives a $0.20/lb premium over the going dock price. The biggest advantage that controlled atmosphere storage provides Hurst's and its customers is the capability to provide high quality fruit even if there is a short-term supply disruption.[7] In 2009 extreme heat resulted in exactly that sort of disruption, and Hurst's was able to use controlled atmosphere stored product to maintain promised deliveries. The retailers who receive the guarantee of an uninterrupted supply of higher quality product have to pay for this assurance. The controlled atmosphere storage also allows Hurst's to continue shipping some Oregon berries for nearly a month after field harvesting ends. The market advantage this provides is somewhat limited because there are Washington production areas that can also supply berries during this late season period. The two local chains have not invested in controlled atmosphere storage and therefore do not have this ability to overcome temporary shortages caused by weather.

In addition to controlled atmosphere storage, Hurst's uses a variety of strategies to extend the Oregon season on both ends, including varietal selection and contracts with growers who are able to fill specific market windows. The latter involves purchasing berries from growers located as far away as Roseburg, Oregon (177 miles south of Portland), in areas with a different climate and relatively little commercial blueberry production.

Producing and packing Oregon blueberries was the initial focus of the business, and even as Hurst's has developed into a year-round supplier, the Oregon product continues to be the most important business component. The packing and distribution division of Hurst's requires a full-time, year-round staff of 25 with 10 of those in the office and 15 in the packing facility. The packing staff moves with the processing equipment when the line moves to California. The sales staff has varied over time but generally consists of two sales representatives and two support staff. As many as 200 seasonal workers are employed in the main packing facility for the period late June through early October.

The total volume of berries that Hurst's distributes is highest during the peak Oregon production months of July and August and stabilizes at a much lower level for the rest of the year. In the shipping process, the berries change ownership at the company dock. Hurst's formerly used its own trucks and staff but eventually decided that insurance and other concerns made it too much to handle in-house. Now the company arranges for transport and bills the buyer for the cost.

The Oregon berries packed by Hurst's have broad distribution in the western United States and beyond. The company estimates that about

15 percent of these berries stay in Oregon. Overall, the Hurst's brand blueberries have a relatively low consumer profile. A comparison of Hurst's web site with sites for Driscoll's (driscolls.com/) and Naturipe (naturipefarms.com/) shows that the latter two have a much stronger consumer orientation, while the Hurst's site clearly targets an industry audience. Allfoods does spotlight Hurst's berries in the Portland market, as the retailer featured Mark Hurst in both print and point of sale ads during the Northwest growing season. Still, many consumers who purchase Hurst's berries from Allfoods undoubtedly think of them as coming from a specific store (Allfoods) rather than from a specific packer (Hurst's), despite the label on the package and the photo next to the display.

As depicted in figure 8, Hurst's sells to four types of buyers—retailers, wholesalers, food service, and exporters. Approximately 50 percent of the Oregon-produced Hurst's blueberries are sold to major national retailers. Each of these retailers now operates in a similar fashion, with a single buyer assigned to purchase blueberries for the entire country on a year-round basis. The range of additional products handled by this buyer varies only slightly from firm to firm. Using Allfoods as an example, the buyer also covers blackberries and raspberries but does not handle the purchases of certified organic berries.

This national, highly concentrated buying approach adopted by the major supermarkets provides two advantages to large scale packer-shippers such as Hurst's. First, the system makes it difficult for smaller-scale producers to enter these distribution channels because the buyer prefers dealing with a packer-shipper who can provide product for a longer period and a broader geographic region. Second, this concentrated approach reduces Hurst's transaction costs. While the Hurst's marketing staff must be in frequent contact with each buyer (multiple times a day), this still represents a large reduction in contacts and accounts when compared to retailers using a regional format. On the negative side, this concentration greatly increases the stakes of getting everything absolutely right, as a single slip-up or perceived problem could result in the loss of all sales to a national client.

About 25 percent of the berries go to wholesalers, who take ownership of the berries and then redistribute to smaller grocers and to other customers, including some food service firms. Direct deliveries to food service accounts represent only 5 percent or less of the business but are viewed as a potential growth area. Finally, Hurst's exports to Europe and Asia by air represent 20 percent of sales. The firm is a leader in this area, considering that only a much smaller percentage of the overall fresh U.S. crop goes to these markets.

During the Oregon production season, grower prices are fairly standard among the different Northwest packer-shippers. These dock or purchase prices are determined in a marketplace that approximates perfect competition. According to the Oregon Blueberry Commission (oregonblueberry .com/), there are 300 Oregon growers seeking to sell and more than 10 fresh market buyers who must take into account supply and demand conditions beyond Oregon. Blueberry prices have varied considerably over the years, but high prices from 2005 to 2007 and substantial media attention resulted in a surge of new plantings. As has happened in the past, grower prices and returns plunged in 2009, although some in the industry view this as a favorable sign, with the lower prices a needed means of encouraging more demand. Prices rebounded considerably in 2010.

At the height of Northwest fresh season in August 2009, the dock price of 80 cents per pound was low enough that all market observers indicated growers were losing money, and some were even leaving berries in the field. A year earlier, the dock price was $1.20. Hurst's and other similar packer-shippers in the Northwest were faring better because some of their services were provided on a fixed cost per unit and therefore did not decline with the price level. On a year over year basis, Hurst's calculated that the company's per unit blueberry revenues were down 12 percent from 2009 as compared to the producers, who were down 33 percent.

Retail Sales

Allfoods, with multiple stores in metro Portland and more than 1,000 stores in the United States, is typical of the major retail stores that carry Hurst's blueberries. While the firm has fruit merchandisers who set margins and design sales campaigns in the Northwest and other regions, a single individual handles all non-organic blueberry purchases. Although there are occasional periods in the fall and spring when blueberries are relatively scarce, Allfoods seeks to have fresh blueberries for sale during the entire year. Allfoods is a self-distributing chain, so the fresh blueberries pass through a Portland Allfoods distribution center before going out to the individual stores.

The Allfoods berry buyer stresses that "local deals are big for us" and that local suppliers receive preferential treatment. The Allfoods annual report states that on a national basis more than 30 percent of the fresh produce sold is local, but it provides no data or definitions. In the Northwest, blueberry purchasing behavior has become more local; there were periods in the past when virtually all the blueberries sold by Allfoods during the height of the Oregon-Washington season were sourced in British Columbia, Canada. In 2009 Hurst's was a featured local supplier, and Mark Hurst's photo was prominently displayed in the produce departments and in print advertising. Still it was confusing for consumers to understand where blueberries were actually produced, as Hurst's products, sourced from both California and Oregon, use the same labels and list the farm headquarters as Sheridan, Oregon.

As is true with other major customers, Hurst's provides special packaging and works with Allfoods to provide products for special campaigns. During the 2009 Oregon production season, Northwest blueberries were frequently featured by mainstream Portland retailers with pack sizes as large as five pounds and prices as low as $1.58/lb. Often sales prices were only 50 percent of the prices charged the previous week. At the sale prices, blueberries were a loss leader earning barely what the retailer paid for the fruit.

Food Safety Considerations

Allfoods, Hurst's, and other participants in the mainstream supply chain have emphasized food safety as an issue of growing importance to the long-run strength of the industry. The strict food safety requirements represent a key difference between the mainstream distribution chain and the other supply chains examined in this research, and they act as a barrier to entry into mainstream distribution for many smaller growers. Large-scale growers believe that for the well-being of the overall blueberry industry, all growers should meet the same food safety standards. As one said, "Food-borne pathogens do not discriminate based on the size of the farm and the size of the processor" (Lies 2010). The 2010 FDA Food Safety Modernization Act (U.S. Government Printing Office 2011) should not result in any major changes in how the mainstream blueberry supply chain operates.

The systems that Hurst's and other Allfoods suppliers have in place allow for traceability of each blueberry package back to an individual farm and specific harvest and packing days. Currently Hurst's uses this information solely as a food safety tracking mechanism and does not make any of it accessible to consumers. Other packers, however, have begun to label their berries with source of production information, and Hurst's may do so in the future. In this way the food safety information could also serve to market the origin of the product.

Hurst's requires the Oregon farms that supply it with raw product to have Good Agricultural Practices (GAP) certification through the Oregon Department of Agriculture (oregon.gov/ODA). In addition, Hurst's own blueberry acres are Global GAP certified (globalgap.org), so the berries can be exported to Europe. The firm's packing facilities have an additional Primus certification (primuslabs.com), as required for products sold to certain retailers. Hurst's has a full-time staff person who focuses on food safety issues and was originally hired to work in Mexico. She now rotates back to the United States to work on the same issues when the Mexican crop is finished.

Food Miles and Transportation Fuel Use

The Hurst's berries sold in Portland area Allfoods stores are transported three times–35 miles from field to packing facility, 50 miles from the packing facility to the Allfoods regional distribution center, and 20 miles from the regional distribution center to the individual store. Thus the food miles or total number of miles traveled is 115 miles.

To calculate fuel use the assumption is made that the distance should be doubled to reflect a round trip with the return empty. The first trip is made in an open truck with a 10,000-pound capacity and a fuel efficiency of 10 miles per gallon. The other two trips are made in 48-foot trucks, with a 40,000-pound capacity and fuel efficiency of six mpg.[8] Therefore the total fuel usage is 0.07 gallons per 100 pounds of fruit for the first trip, 0.04 gallons per 100 pounds for the second trip, and 0.017 gallons per 100 pounds of fruit for the trip from the distribution center to the individual stores. The total fuel used is thus 0.127 gallons per 100 pounds of fruit for transport from the field to a Portland area retail store.

Community and Economic Linkages

Allfoods seeks to be a good corporate citizen in the Portland area and has wide involvement in many activities that support diverse health, youth, and other charitable organizations. Relatively few are local food–related. Mark Hurst and his firm play key roles in the state and national blueberry industries. The firm has limited involvement in local food system initiatives. Both Allfoods and Hurst's contribute high levels of wage income to the local economy, as their respective industries are quite labor intensive.

Prospects for Expansion

In the very competitive supermarket industry, Allfoods remains a key and growing player. It is one of the firms that has benefited from the growing concentration of that industry. Allfoods plans to take full advantage of Portland's proximity to commercial blueberry production by continuing to promote in-season local blueberry sales.

As a major player in the blueberry industry at the national level, Mark Hurst is well aware of product trends and consumer research, and he believes that fresh blueberry markets are not yet saturated and have potential for growth. Within that expanding marketplace, he regards the size of his own firm as both an advantage and disadvantage. On the plus side the firm is small enough to turn on a dime and provide what each individual customer needs. Hurst's provides control all the way to the retail store. But size is also a disadvantage, as he must compete with larger firms that have more resources and a higher profile in the market place. Thus the possibility always looms that he could lose either key clients or key growers.

Still Hurst remains intrigued by the possibilities of gaining new success in a very competitive market "by figuring new things out." This is why he pioneered blueberry production in Mexico, where he is experimenting with growing berries in several different climatic zones. So far that production is sold domestically in Mexico, but in the future it might be exported to the United States.

Key Lessons

There are three key lessons learned from this examination of the mainstream supply chain. First, the broader marketplace drives participants at each level to become ever more reliable and efficient. This has resulted in increasing consolidation and coordination. Packer-shippers recognize that they must develop the capacity to source product year-round and must stay on the cutting edge in terms of storage, packaging, and transportation. In the future, machine harvesting of fresh berries will likely be required to remain competitive in the mainstream supply chain. These changes and new food safety requirements make it difficult for smaller producers to enter these mainstream chains.

Second, although the tools are in place to provide consumers with more information about where their berries are grown and who produced them, this information is not commonly provided to consumers. It remains unclear how and when the mainstream supply chain participants will begin to communicate this information to consumers.

Third, the blueberry industry recognizes a need to continue to focus on strategies to increase demand and move more product. Consumers have demonstrated a willingness to buy larger containers if per pound prices are lower, so it is likely that they will continue to find this option in supermarkets.

Direct Market Supply Chain: Thompson Farms

Thompson Farms, located in Damascus, Oregon, on the urban fringe of Portland, was established more than 60 years ago when Victor and Betty Thompson began growing strawberries. In the 1980s their son Larry was farming part-time and selling his three crops (broccoli, strawberries, and raspberries) primarily to processors. After considering his alternatives, Larry Thompson concluded that given his limited acreage and his proximity to urban markets, he could only farm full-time successfully by refocusing the farm on direct market supply chains that would produce greater per acre returns. Currently the farm produces 50 berry and vegetable crops that Thompson direct markets through diverse supply chains to local consumers.

Thompson describes his business as "figuring out what our customers want and providing it" (Lev 2003). This simply stated objective has led Thompson through numerous changes over the years and the development of complex production and marketing programs. Marketing rather than production remains the driving force. A farm profile written as a part of a 1989 study characterizes Thompson as " a clever marketer, but not as up-to-date on his production program as he could be" (Lev 2003). This continues to be true in 2009.

In searching for a direct market case for this study, the preferred alternative of a blueberry-focused full-time farmer who sells exclusively through direct markets proved difficult to identify. Three options were considered: (1) a part-time farmer with a strong blueberry and direct market focus, (2) a full-time farmer with a focus on blueberries and a mix of direct and other marketing approaches, or (3) a full-time farmer who sells exclusively through local, direct market chains but has less focus on blueberries. Thompson Farms fits in the third category and was chosen because it provides the best opportunity to develop an understanding of Portland area direct markets. Thompson Farms uses three distinct outlets for the crops produced—farmers markets in the Portland metro area, permanent farm stands, and temporary farm stands set up on the grounds of Portland area hospitals.

Supply Chain Structure and Size

Figure 9 shows the diverse market channels used by Thompson Farms. As detailed in the following section, all products are direct marketed and all are sold within 20 miles of the farm.

Production

Thompson Farms' diversification of crops and market outlets is typical for full-time direct market farms in the Northwest. Berries (strawberries, raspberries, boysenberries, Marion blackberries, and loganberries in addition to blueberries) represent 60 to 65 percent of the farm's total revenues. Blueberries are produced on seven of the 145 acres, or 5 percent of the acreage. Deriving from a relatively high value crop, blueberry sales represent a somewhat greater share of the farm's total revenue. All of the blueberry acreage has been in production for "a long time,"

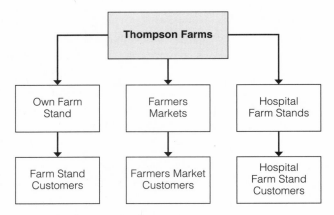

Fig. 9. Direct market supply chain, Thompson Farms.

and Thompson has no current plans either to plant additional acreage or to switch varieties. He indicated that he would consider a switch if significantly better varieties came along.

Farm-direct production and marketing are labor-intensive activities. Thompson Farms has four full-time year-round employees, six additional field production employees (March through October or November), three seasonal packers who prepare produce to go out to the markets and farms stands, and a seasonal sales staff of 12 who work 38–40 hours a week during the main production season. The field employees are all Mexican nationals who have remained with the farm for more than 20 years. Thompson believes that recruiting and managing staff is one of the strengths of his business.

The Thompson Farms blueberry production practices are roughly similar to those of other Oregon fresh market growers as described in a 2005 enterprise budget (Eleveld et al. 2005). The farm's estimated yields of 5,000 pounds per acre are only about 60 percent of the average yield reported by USDA for Oregon blueberries (USDA-NASS 2008). While a part of this shortfall reflects the reduced yields from fresh as compared to processed production, Thompson views his crop diversification as the major reason for his below average yields; he simply does not stay on top of his blueberry crop to the same extent that more specialized producers do.

The basic production process, as detailed in the enterprise budget, requires expenditures for irrigation, the application of sawdust every third year, pruning, and hand harvest. As is almost always true for fresh market blueberries in Oregon, Thompson Farms blueberries are hand harvested, and those costs represent about 60 percent of the total production cost. While less expensive machine harvesting may be in the future for this crop, very few producers have made the switch for fresh market blueberries. At Thompson Farms the field workers harvest the blueberries into large buckets and then, while still in the field, transfer the berries into one-pint containers. These containers are then placed in 12-container flats. Each flat contains 9–10 pounds of fruit. Thus the two units used by Thompson Farms to sell blueberries are dry pints and flats. The berries are brought back and cooled down in a walk-in refrigerator before they are transported to the different market outlets.

Thompson has long focused on using "gentle" farming techniques, such as cover cropping and holistic pest management strategies, and advertises all his products as "no-spray" and/or "insecticide and fungicide free." In an earlier period he was one of the first growers certified by the Portland-based Food Alliance (foodalliance.org), but he subsequently gave up that certification as he judged that given his target markets, it was not worth the time and cost. Thompson Farms products currently have no third party certification.

Because he targets direct markets, Thompson seeks to have a longer production season as compared to growers who are harvesting either for fresh market packer-shippers or for processors. In 2009 he sold berries in farmers markets for nine weeks. Blueberries are available through his farm stands beyond his own production season because he resells berries for a farmer who has fields with a harvest period that extends later into the fall.

Marketing and Distribution

All the Thompson Farms blueberries are sold in the Portland area and within 20 miles of his farm. Over time Thompson has experimented with many different Portland area direct channels, and ultimately his selection of marketing outlets drove the changes in farm production. His current distribution scheme is split into three parts, with approximately 30 percent sold through his three farm stands, 35 percent through seven farmers markets, and 35 percent through seven periodic farm stands hosted on hospital campuses. He also has a small U-pick operation for strawberries and pumpkins but not for blueberries. Thompson recognizes that distribution channels go in and out of favor, and therefore he remains willing to change. For example, nearly 20 years ago he went through a period in which he ran very successful farm stands in the parking lots of a national supermarket chain. At a certain point, Thompson and the store managers made a mutual decision to end the relationship. More recently he gave serious consideration to starting a community supported agriculture program but ultimately decided against it.

All these outlets allow the farm's sales staff, using carefully crafted signage and discussions with customers, to highlight where and by whom the crops have been produced. Thompson believes that freshness and flavor are the two characteristics that consumers value most, and he therefore minimizes the time from the field to sales locations. For all market outlets, berries are directly harvested into pint containers and then transported to the staging area, where the products are refrigerated. In most instances the berries leave the farm the next day to be delivered to one of the market outlets (a single delivery truck makes the initial rounds and then resupplies a farm stand or market as products run short).

FARM STANDS: One of Thompson's own farm stands is on the farm, while the other two are on leased property in neighboring towns. All three are simple structures more similar to a farmers market booth than an elaborate enclosed farm stand. Thompson Farms does not operate a web site, although these farm stands are listed on web sites developed by third parties. Thompson does have a telephone "Crop Update Line" and believes the enthusiastic messages that he records each morning are successful in generating business. Prices at the farm stands are generally 10 percent below the prices offered for the same Thompson products at either the farmers markets or the hospitals. The farm stands also feature frequent deals on quantity purchases of whatever is in abundance. The farm stands, especially the one on the edge of the farm itself, are open for a longer selling season than the other market outlets. They face competition from nearby farm stands but have been quite successful over the years.

FARMERS MARKETS: All of the seven farmers markets that Thompson supplies are in the eastern Portland metro area and are an easy driving distance from his farm. Since there are 40 farmers markets in metro Portland, Thompson has been able to select markets where his products fit well. Thompson Farms does not participate in very early and very late season markets, as Thompson is careful to ensure that he has enough to sell and there will be sufficient customers in the market to make the effort worthwhile.

The Thompson Farms berry crops are sold during the prime market season and during the period when berries attract customers to the booth where many also purchase vegetables. In most markets and through his farm stand, Thompson takes advantage of rules that allow him to bring in peaches from a partner farm in eastern Washington. In Portland direct market circles, it is well known that "sugar crops" such as berries and peaches drive customer traffic.

Although most markets place some restrictions on the number of vendors with similar products, Thompson Farms still faces significant competition. As an example, during the prime market season, Thompson Farms was one of 15 vendors selling blueberries (10 conventional and five organic) at the largest market he attends. Blueberry prices ranged from $2.67/lb to $4.66/lb during the weeks that Thompson farms sold berries in the 2009 season, with most conventional berries sold at $3.33/lb to $4.00/lb. Quantity sales of half-flats (six pints) and flats (12 pints) were sold at a 17 percent discount. Thompson set his own prices for each market day, so in some weeks he was lower than the competition and in others he was not.

Farmers markets are high stress environments for producers as the mix of growers and the rules change frequently. Thompson frequently served on market boards in the past but has chosen to step back as he was often frustrated by his participation. He remains concerned that market rules prohibiting resale of wholesale products are not being enforced, meaning that farmers who follow the rules face unfair competition. On the other hand, he freely admits that his large farm size (as compared to other market vendors) allows him to price "specials" aggressively. Blueberries are not generally a featured product at his booth, but some of the other berries, such as Marion blackberries, often are. Overall, market sales have been stable in recent years, and this is not a growth area for the business.

HOSPITAL FARM STANDS: Several years ago, through a connection with a Kaiser-Permanente consultant, Thompson developed a new distribution channel—farm stands hosted by hospitals. While Kaiser organizes and/or sponsors many farmers markets, in certain instances a single farm stand represents a less disruptive and therefore preferable approach for promoting farm direct marketing on a hospital campus. For Thompson Farms the exclusive access to a population eager to buy fresh produce provides an excellent and profitable opportunity. According to a Kaiser spokesperson, "We treat the farm stand the way we would our campus pharmacy. When our physicians prescribe a healthier diet for patients, the remedy is right there at hand" (Western SARE 2008: 6). Blueberries, because they are viewed favorably by the medical community, are quite popular at these stands.

Subsequently Thompson Farms expanded the number of days at the first hospital and added a second hospital based on a connection through a neighbor. By 2009 his sales at hospitals had expanded to seven days, two days each at two hospitals and a single day at three other hospitals. He also sells in an actual farmers market located on a hospital campus, so overall he has developed a strong relationship with the Portland medical community. The gradual development of this new sales channel demonstrates Thompson's incremental experimentation playing a key role in keeping the farm profitable in recent years by allowing him to sell more product. The prices charged at these sites are the same as the prices charged in farmers markets.

Food Safety

As a function of the outlets used and the products sold, Thompson Farms has not had to undergo food safety inspections. In fact, since Thompson left the Food Alliance program, the farm's production no longer undergoes any third party scrutiny. If food safety certification became mandatory for farmers who direct market, Thompson Farms would have to absorb this additional expense. However, because he is a relatively large direct marketer, such a regulatory change might actually provide the farm with a competitive advantage as compared to smaller farms. It appears that Thompson Farms will be subject to the provisions of the FDA Food Safety Modernization Act because the farm's sales exceed the small farm direct-market exemption.

Food Miles and Transportation Fuel Use

As described, Thompson sells blueberries through three channels— traditional farm stands, farmers markets, and farm stands hosted by hospitals. Across all three of these channels, the maximum distance is 20 miles from where the blueberries are produced to where they are sold. The average distance is 10 miles and must be doubled to reflect the return trip, resulting in 20 roundtrip truck miles. Based on transporting 1,500 pounds of product an average roundtrip distance of 20 miles in a panel truck with an 11 mpg fuel efficiency rate, 100 pounds of blueberries require 0.12 gallons of fuel.

Community and Economic Linkages

Larry Thompson has been active in the sustainable agriculture movement for the last 25 years and was the first farmer to serve as board president of the Western Sustainable Agriculture Research and Education program. In 2008 he was awarded the national Patrick Madden Award for Sustainable Agriculture in recognition of the breadth and quantity of his achievements.

Over this period he has freely shared his insights and expertise through numerous conferences, workshops, and farm visits and has been the subject of several previous case studies. His stated goal is to increase public understanding of urban fringe agriculture: "Instead of seeing my farm as a secluded hideaway, I am getting the community involved, bringing them to see our principles in action" (USDA-NIFA 2009). Currently six immigrant farmers working with Mercy Corps Northwest are farming on 3.5 acres of his land and selling their products through his farm stands. He is intrigued by the possibilities for expanding these sorts of mentoring efforts.

Thompson Farms is a labor-intensive enterprise with a substantial wage bill. Thompson tries to ensure that his field workers earn between $15 and $25/hour throughout the year. The packers earn minimum wage, and the sales staff earn minimum wage plus bonuses. As a direct marketer of an unprocessed product, the farm pays fees for using space at the farmers markets and one of the hospitals, but these fees are fixed and quite low.

Prospects for Expansion

Over the last 25 years Thompson Farms has demonstrated the ability to carve out a profitable but changing niche in local food markets. The relatively stable direct market prices received for blueberries have allowed Thompson to avoid the boom or bust prices received by mainstream growers in the commodity blueberry market. Unpredictable prices in processing markets were the primary motivation for Thompson to switch over to direct marketing more than two decades ago.

While direct marketing has greatly expanded in Oregon, so have the number of farms active in this sector. In the face of this new competition, Thompson Farms has continued to do well. Thompson's son is considering joining the business full-time, which has caused the family to consider expansion options. Some urban fringe farmland continues to be available to lease or purchase, and complementary enterprises such as cattle production for on-farm sales are possible. Thompson believes, however, that the implementation of a mixed land use system that would allow him to sell some of his existing acreage selectively for development, while continuing to farm more intensively on the remaining acres, will be critical to the long-run viability of full-time urban fringe farms (Meunier 2009). In the last few years this has led him to become an active participant in the land use planning process at the local and state levels and in a variety of research efforts.

Key Lessons

Three general lessons emerge from this direct marketing case study. First, the decision to rely solely on direct marketing outlets has forced Thompson Farms, a full-time farming operation, to grow a diversity of crops. If Thompson Farms wanted to focus solely on blueberries or any other narrow crop category, the farm would have to sell through other types of markets in addition to direct markets. In contrast, part-time farms do exist in the Portland area that produce only blueberries and sell only directly.

Second, this case demonstrates that the profitability of direct market outlets changes over time, and farms must be willing to respond.

Thompson regards his flexibility as a key strength. His expressed farm goal is to meet customer demand rather than supply any specific market outlet or set of outlets.

Third, land use rules will have a major impact on the long-run viability of urban fringe farms similar to this one. Thompson believes that more flexible land use rules are essential for keeping farms such as his profitable in the future.

Intermediated Supply Chain: New Seasons Market and Nature's Fountain Farm

Although substantial quantities of blueberries are sold through farmers markets, farm stands, and U-pick operations, many Oregon consumers do not shop at these direct market channels.[9] The sale of identity-preserved, local products through mainstream retailers provides consumers a convenient alternative way to purchase these products. The integration of one or more links (or intermediaries) into a local supply chain is termed an intermediated distribution chain.[10]

This case study examines two participants in the metro Portland supply chain for local blueberries—a retailer and a farmer. It also briefly examines the role that a third participant, a distributor, could play.

New Seasons Market, a chain of ten Portland supermarkets founded in 2000, prides itself on being different (newseasonsmarket.com/). In a 2006 *New York Times* profile of the company, Phil Lempert from Super marketguru.com gave this assessment (Burros 2006): "The New Seasons model is a brilliant concept because it brings back the days of food co-ops, the feeling of being closer to nature, to the food supply, to the neighborhood. What they are saying is, we are your store and we want to build a relationship with you. That lack of relationship has been the downfall of supermarkets."

In the Portland market New Seasons must compete with a broad range of competitors, including national supermarket chains, national and local natural food stores (including consumer cooperatives), farmers markets, and farm stands. The key strengths that New Seasons emphasizes are the scope and quality of its "home-grown" offerings, excellent in-store service, and active participation in and support of community activities.

New Seasons purchases blueberries from about a dozen Northwest producers but works intensively with a much smaller number. Scott Frost, the owner of Nature's Fountain, has been one of New Seasons' main suppliers of organic blueberries since the chain opened. Frost has been involved in farming in one way or another for 35 years and has farmed at his current location, 60 miles south of Portland, since 1996. The farm consists of 26 acres, of which 15 acres are farmed. Half of the producing land, 7.5 acres, is in blueberries. So while Frost grows more than 50 crops, through the 2009 growing season the farm had a strong blueberry focus.

Although New Seasons accepts direct deliveries, it also uses a variety of produce distributors. Its preferred distributor, Organically Grown Company (organicgrown.com), has a well-developed program to work with Northwest growers. Inserting a distributor in this local supply chain would introduce both advantages and disadvantages.

Supply Chain Structure, Size, and Performance

Figure 10 depicts the intermediated food supply chain that allows Nature's Fountain to sell its blueberries to New Seasons Market customers.

Production

The permanent farm staff for Nature's Fountain consists of Frost, his wife Sylvia Cuesta (who is somewhat less involved with the farm right now because of off-farm work), and two full-time employees. While in the past Frost hired as many as 20 workers for the six- to eight-week blueberry harvest period, in 2009 he hired only eight. In 1998 Frost was the first Oregon agricultural producer to sign a contract with a farm workers' union. Eventually that agreement fell through, as Frost felt that the unionized workers were making less with a fixed hourly wage than they would earn from harvesting blueberries for a per pound rate (Northwest Labor Press 1998).

The blueberry acres (and his entire farm) have been organically certified for many years, but Frost decided to let the certification lapse, although he continues to farm in the same way. He does not believe that the organic certification process achieves its original intent and does not feel it is worth the time and expense.

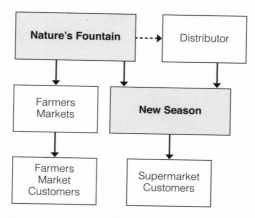

Fig. 10. Intermediated supply chain, Nature's Fountain and New Seasons.

Frost estimates his yields at 5,000 lb per acre, or about 60 percent of the reported statewide yield of 8,000 lb/acre (USDA-NASS 2008), and recognizes that his production methods do not maximize yield. According to the most recent Oregon State University blueberry enterprise budget, more than 60 percent of the production expenses for fresh market, hand-harvested blueberries are the harvest costs (Eleveld et al. 2005). Nature's Fountain field workers earn $0.45/lb to $0.50/lb (depending on whether the worker stays for the entire season) to harvest the blueberries into large plastic buckets. The buckets are brought to a sorting shed, where four workers in a separate group clean them on a simple (and quite old) mechanical sorting line. Frost estimates that the sorting costs are an additional $0.40/lb and views this step in the process as one he needs to think about changing in order to reduce his costs. He knows that his total harvest, cleaning, and packing costs of $0.85/lb to $0.90/lb exceed what other producers spend on these functions. The approach followed by Thompson Farms in the direct marketing case, which does not include a sorting line, is more typical of what other local small-scale blueberry farms are doing.

Marketing and Distribution

In 2009, 50 percent of Nature's Fountain's production went to New Seasons markets in metro Portland and 50 percent was sold through three

Portland area farmers markets (Moreland, Lake Oswego, and King). The markets are medium sized with managers and management styles that appeal to Frost. For both retail outlets and farmers markets, he has found that he does better traveling to metro Portland rather than selling in the nearby midsized towns of Corvallis and Albany. As discussed in the section on community linkages, this decision to sell in more distant metro markets has an impact on his ability to become involved in community activities.

While he sells only blueberries to New Seasons, Frost's farmers market booth features a broad range of 50 heirloom vegetables that he also sells to a limited number of chefs in Portland. His vegetable crops are more differentiated from other producers' vegetables than are his blueberries.

Although Frost hires labor to work on the farm, he personally handles the marketing with some help from family members, especially his wife. At markets, Frost describes his presence and approach as follows (Meyers n.d.): "Don't ask me a question if you don't have some time. What I do is much more than a cash exchange. I wish everyone could have the experience of growing their own food, even if it's just a small window garden. But if you can't, guys like me will do it for you. And I will show up at the market religiously to provide that service." His market booth is more diversified than the majority of market booths but is still smaller than the large "anchor" booths at Portland markets. Frost has carved out a market niche between the small growers who have a narrower range of products and the large growers who offer less personalized service and generally staff their booths with employees who do not participate in production activities. Frost's approach, however, limits the time he has available to work on-farm and also does not allow him to sell at multiple markets on the busiest market day, Saturday.

His early and late season blueberries, available in smaller quantities, go exclusively to the markets rather than New Seasons, for three reasons: (1) he would not be able to meet the demands of all the New Seasons stores, (2) he can earn full retail for the berries at the higher early and late season farmers market prices, and (3) selling the berries at his stand attracts customers who then also purchase additional items. In midseason, however, New Seasons represents his primary blueberry sales outlet by volume.

The New Seasons produce departments strive to present a farmers market–like image as the produce managers seek to identify all products by production location and, when appropriate, to provide individual farm names. During the local growing season, produce stands are set up in the parking lot on selected days to feature specific local producers.

Produce purchasing is centralized through a two-person department located at the corporate headquarters. Jeff Fairchild, the head produce buyer, recognizes that working effectively with local growers is a complex, long-term process. The first year is a small-scale trial period spent figuring out how to work together so that a more effective relationship can be established for future years. Given the number of farms from which New Seasons buys, and the time required to develop strong and successful relationships, Fairchild restricts the number of new farm suppliers in a given year. He also consciously limits the number of suppliers per produce item rather than seeking to maximize competition among growers.

Fairchild sees buying from smaller local growers as both enjoyable and consistent with New Seasons' mission and believes this will continue to be the way they do business. But he also recognizes that pricing pressures may play an increasing role in sourcing decisions, and this has already begun to drive him to buy from larger-scale, lower-cost growers. As market conditions change, one of Fairchild's main goals is to be honest with his suppliers about what this will mean for New Seasons' price and quantity requirements. In his view, if business relationships are no longer working well for both sides, it is best to get people prepared for the changes they will need to make.

At about 0.5 percent of total produce sales, blueberries are a small but growing category. As is true with mainstream supermarkets, New Seasons carries blueberries throughout the year, sometimes as a featured item. Fairchild estimates that Northwest-produced fresh blueberries are available for 10 weeks and that these berries represent 25 percent of the berries sold by New Seasons for the year. Oregon berries, all purchased from individual farms, are available for only about half of this period, or five weeks on average. Some of the Washington blueberries are from relatively small farms, while other Washington berries are distributed by large packer-shippers who source in that state. When it comes to geography, Fairchild believes that New Seasons customers do not distinguish between Oregon and Washington because both are viewed as local blueberry sources.[11]

When asked about New Seasons' strongest competition during the local blueberry season, Fairchild first mentions farmers markets but also indicates that he must pay attention to what other retailers are charging. One trend he has noticed is that the national chains are aggressively pricing blueberries in larger pack sizes.

New Seasons has purchased in-season berries from Scott Frost since the chain opened in 2000. They have no formal contract but simply pick up where they were the previous season. Fairchild telephones Frost to place orders for all the stores.

The individual stores have limited produce storage capacity, and New Seasons has chosen not to operate its own distribution center. As Fairchild explained when interviewed for this study, "Having a warehouse probably is not a business function that I will want to take on even as New Seasons expands." Individual stores receive produce either directly delivered by farmers or as a part of a distributor delivery. In general, Fairchild prefers using a distributor, even for local products, explaining, "Sometimes people go overboard when they stress the importance of maximizing direct links to farmers. I think farmers should have the time to farm rather than having to run around to all nine of our stores." Exceptions are made for particularly fragile, short shelf-life products, such as most berries. Blueberries, however, probably do not need to be in that category. A web site focused on the at-home shelf life for consumer products (stilltasty.com) gives blueberries a stored life span of 1–2 weeks compared to 2–3 days for blackberries and 3–5 days for strawberries.

Nature's Fountain blueberries have always been in the direct delivery category. Frost himself makes the deliveries to the New Seasons stores and requires nearly a full day to make the rounds. As an alternative in 2009, Frost experimented with using New Seasons preferred distributor (Organically Grown Company), but the two did not work well together in terms of logistics, and Frost was reluctant to pay the 10 to 12 percent handling fee charged by the distributor.

The individual store produce managers are quite happy to receive direct shipments from growers such as Frost, but that is probably because the number of producers making their own deliveries is carefully managed.

A dramatic increase in direct store deliveries would greatly add to the management challenges at the individual stores.

At New Seasons, the produce department provides customers with a visual cue for the arrival of local blueberries by selling these berries in open cardboard pints as opposed to closed plastic clamshells. This provides a farmers market feel to the product and an impression of freshness. Still, even during the heart of the local season, the larger pack sizes of local berries are sold in plastic clamshells.

Pricing is a significant challenge in this intermediated, direct-to-retail chain. In the mainstream chain, prices are determined by the broader blueberry market. In the direct markets, producers have some ability to set their own prices. In this chain, while the actors value their relationships, they also recognize both the need for and the challenge of determining a mutually beneficial price. This proved difficult for New Seasons and Frost in 2009 and has since precipitated a change.

Fairchild is constrained by competing prices in the marketplace and therefore needs to set purchase prices low enough to earn an adequate margin for New Seasons. His general goal for the produce department is a margin approaching 50 percent of the retail price. Particularly for local products, this is not always achieved. Frost, however, wants to charge New Seasons the same discount price (17 percent less than the retail price) he has established for bulk farmers market sales. In his mind this is what the berries are worth and therefore the minimum acceptable price. The related idea of a "dignity price" for a product has been discussed by other growers (see, for example, Stevenson 2009, for the 2008 case study of Red Tomato, a northeast U.S. produce distributor).

A wholesale price set at that level (retail minus 17 percent) would make it impossible for New Seasons to earn a 50 percent retail margin and still charge prices that are similar to farmers market prices and competitive with other retailers. The conflicting nature of these price goals was heightened in 2009 by the large national and Northwest blueberry crops, which led to reduced mainstream blueberry price levels. With some discomfort, Fairchild ended up paying something close to Frost's minimum price requirement, earning a reduced margin and still setting the New Season retail price (about $5.45/lb) higher than the farmers market price for organic berries for that time period.

For five of the 10 weeks that make up the local season, New Seasons also offered a substantially larger pack size, in either two or three pounds, that was supplied by a local producer other than Nature's Fountain. These larger packs provided the lowest price offered by New Seasons, $3.33/lb, and provided additional competition for Nature's Fountain pint sales.

Food Safety

Food safety policies may also influence the types of farms that will supply New Seasons and similar stores in the future. In 2009, in contrast to mainstream supermarket chains, neither New Seasons nor Organically Grown Company required farm suppliers to have any third party food safety certification. Both firms indicate that one key reason behind this decision was the recognition that certification costs would place a relatively greater burden on smaller growers. Fairchild's 2009 prediction that within two to three years some sort of certification would be required turned out to be true with the passage of the 2010 FDA Food Safety Modernization Act. Nature's Fountain appears likely to fall into the exempt category, with sales below the $500,000 sales level and all sales within state. The farm does not have any food safety certification and, as mentioned earlier, is planning to drop its organic certification because of cost concerns. Frost has been outspoken in his opposition to proposals to mandate food safety certification requirements for growers of all sizes, arguing: "All farmers are for food safety, but this [mandated certification] could make farmers markets go away. The only guys left standing in the room will be the big gorillas" (Terry 2010). His reasoning is mirrored in the new legislation.

Food Miles and Transportation Fuel Use

The Nature's Fountain blueberries sold by New Seasons have traveled an average of 70 miles from where they are produced as Frost travels in a big loop to make deliveries to the individual stores. The roundtrip distance is 140 miles since he returns with an empty van. The products are transported in a delivery van with a capacity to transport 1,800 pounds of fruit per trip.[12] The van has a 13 mpg fuel efficiency rate, so the fuel use per 100 pounds of fruit is 0.5 gallons.

Community and Economic Linkages

New Seasons has a strong community program centered on donating 10 percent of after-tax profits to community groups and activities. The firm specifically targets its support to local food systems activities and initiatives. The New Seasons web site lists eighteen farmers markets that have received contributions and provides a complete listing of local U-pick and farm stands. All these outlets, it should be noted, are in some sense competitors with the New Seasons produce departments. In addition, through newsletters and blogs, New Seasons management tries to maintain a high level of dialogue with customers, especially on food and agriculture related topics. In Portland, New Seasons is regarded as an outstanding employer that offers its employees relatively high wages for the retail food industry, health care, and profit sharing.

Frost farms in a rural community with limited market outlets and therefore must travel to an urban area to sell his products. This restricts his community involvement in both places. He has offered organic gardening classes in the neighboring town of Albany but with only limited success. Although virtually all of what he produces is sold in Portland, he does not feel much connection to that community beyond his participation in the three farmers markets. As a fresh market producer, Frost pays significant wages to produce and harvest his many products.

Prospects for Expansion

The growth prospects for local blueberries in natural food stores such as New Seasons continue to be bright as both the characteristic "local" and the blueberry itself are riding waves of popularity. Price pressures within the metro Portland market are likely to result in somewhat larger blueberry producers emerging as the preferred local suppliers. Nature's Fountain will likely scale back blueberry production in favor of increased diversification as the farm moves toward using alternative types of direct markets.[13]

Key Lessons

Three lessons are apparent from this consideration of the intermediated supply chain for local blueberries in Portland. First, this case study demonstrates that the participants in this intermediated supply chain face price pressures from both the mainstream supply chain and direct markets. The result will likely be the retailer seeking out alternative local suppliers who will accept lower wholesale prices for their berries. Frost seems to recognize that he will need to refocus on direct markets.

Second, mandatory food safety certification, either public or private, is likely to have a significant impact on how this supply chain operates. Although smaller producers are exempt from the new food safety legislation if they market most of their sales directly to consumers, they will need to comply if they begin selling mostly to retailers. These new regulations, therefore, do become an additional barrier to participation in this type of supply chain.

Third, this intermediated supply chain case study demonstrates the importance of examining the implications of adding additional links to a local supply chain. Specifically, do consumer perceptions of the localness of the berries decrease as the berries pass through more hands?

Cross-Case Comparisons

The three case studies conducted in Portland illustrate different ways that locally produced blueberries reach area consumers. This final section summarizes similarities and differences across the three case studies and concludes with a brief discussion of key lessons that emerge from these case studies.

Supply Chain Structure

As expected, the three supply chains have quite different structures. By definition, the blueberries do not change hands in the direct market chain (although Larry Thompson does sell blueberries produced by others through his farm stands). In the intermediated chain the berries change hands once, while in the mainstream chain the berries change hands twice before they are sold to the final consumers by Allfoods.

In general, the information transmitted through the local and intermediated chains to consumers exceeds the information provided through the mainstream chain. Larry Thompson's direct market sales provide his customers with an opportunity, through signage and discussion, to understand where, by whom, and how the blueberries were produced.

The scale of his operation is such, however, that consumers are in contact with sales staff, not Thompson himself or others who do the field work. The intermediated food chain producer, Scott Frost, farms on a smaller scale and is much more hands-on with the portion of his crop that is direct marketed. New Seasons, the retail outlet for this intermediated supply chain, labels "home grown" products and also seeks to provide the consumer with information about the source of berries through signage and direct discussions with consumers (the individual produce managers know Frost well enough to represent him).

In contrast, while the mainstream berries distributed by Hurst's Berry Farm and sold at Allfoods do provide some information about where the berries were produced, the information is sometimes misleading. California berries are sold with the Oregon address on the back of the packages, which may incorrectly lead consumers to believe those berries were produced in Oregon. When Oregon-grown berries are sold, Mark Hurst's photo calls attention to the local nature of the product, but the signage may lead consumers to believe the berries were produced by his own farm, when in fact the vast majority are produced by other Oregon farms and packed by Hurst's.

Trading partner relationships are valued and durable in all three of the chains but differ in nature. The mainstream chain relationships are longstanding and stable. These relationships reflect business partners with specific expectations. The direct market chain relationships cannot be easily characterized or summarized. Across the diverse sales outlets used, Thompson Farms sells to hundreds of customers each week. Some have purchased products from the farm for more than 20 years and profess a deep attachment to the farm, while others are hard pressed to remember even the farm name 20 minutes after the purchase. While direct markets imply a special kind of closeness and trust, they do not necessarily include information sharing, decision sharing, and interdependence. The relationship between the intermediated retail produce purchaser (Fairchild) and the producer (Frost), while close and durable, ended following the 2009 season. Even as it became clear that the farm would no longer supply the retailer, both participants characterized the communication as open and straightforward.

Because Portland is within a major commercial blueberry production area, there is some interaction among these three supply chains. The mainstream chain prices are directly linked into the national blueberry market. Still, throughout the year and especially during the local season, prices in different supermarkets may vary considerably because blueberries are often an advertised item. Producer prices in the direct market chain are largely decoupled from the national blueberry market, but the Thompson Farms berries do face considerable local competition. Very few large-scale blueberry producers choose to sell in local direct markets, but that could change in the future and would have significant negative implications for small, locally focused growers. The intermediated chain berries, sold through a regional supermarket, face price pressures from both the broader blueberry market and the local direct markets.

Hurst's, the mainstream grower-packer-shipper, has strong and important links to the wide range of blueberry infrastructure available in the Northwest, including research, education, promotion, packing, and processing.[14] Although neither Thompson Farms nor Nature's Fountain participates to any great extent in blueberry industry activities, they do receive some benefits (especially in terms of knowledge and promotion) from the strength of this industry in the region and the climatic conditions. While their yields are lower than those of large-scale Northwest growers, they exceed the yields achieved by fresh market, locally oriented growers in other regions.[15]

Both the direct and the intermediated chain participants are well integrated within the local food network of other farmers, government agencies, and non-governmental organizations. Thompson Farms developed a series of farm stands on hospital campuses based on relationships formed through his engagement in local food systems activities and is exploring mentoring opportunities for immigrant farmers based on similar ties. New Seasons plays a key role in a whole host of local food initiatives and believes that these activities contribute to its success. The mainstream chain participants, Allfoods and Hurst's, have little interaction with other local food initiatives.

Supply Chain Size and Growth

The three supply chains differ greatly in size. The mainstream supply chain handles the vast majority of Northwest-produced fresh berries and extends far beyond the region, because the quantity produced exceeds the local market demand by a factor of ten. In season, the direct and intermediated market sales do represent a significant proportion of total demand. However, the local season is relatively short, so across an entire year the proportion of local supply in the Portland market is restricted. Lack of year-round availability clearly limits market opportunities for locally produced blueberries. Some berries produced by small local producers do continue to be marketed through local supply chains as frozen or dried products, but this represents only a very small percentage of locally sold products. Neither of the producers featured in the local supply chains chooses to process and sell blueberries in this fashion. The mainstream packer-shipper does use controlled atmosphere storage to extend the marketing season, but the investment cost precludes many small producers from making use of this option.

Entrance into the mainstream chains is difficult and expensive for smaller producers. Both the distributor (Hurst's) and the supermarket (Allfoods) require third party certification, which many small producers such as Thompson Farms and Nature's Fountain do not have and may not be interested in obtaining unless they are forced to do so by law or retailer requirements.

Supply Chain Performance

Even after marketing costs have been subtracted from gross returns, the direct and intermediated chain producers earn much higher per unit returns than do mainstream producers.[16] Table 5 documents the prices and revenues across the three supply chains. The net producer revenues are $2.43/lb and $2.53/lb respectively for the direct and intermediated producers, versus $0.86 for the mainstream producers. The proportion of the revenue retained by the producer varies from a low of 27 percent in the mainstream chain to 46 percent in the intermediated chain and 73 percent in the direct market chain. The retailers in the mainstream and intermediated supply chains earn roughly the same percentage share of the retail revenue.

Aside from organic production methods, there is little product differentiation for blueberries. It is difficult to determine from the price data whether the "local" designation provides a price premium, because

Table 5. Allocation of retail revenue in Portland, Oregon— blueberry chains, by supply chain and segment

| | MAINSTREAM | | DIRECT | | INTERMEDIATED | |
| | *Allfoods* | | *Thompson Farms* | | *New Seasons* | |
Supply chain Segment	*Revenue ($/lb)*	*% of total*	*Revenue ($/lb)*	*% of total*	*Revenue ($/lb)*	*% of total*
Producer	0.86	26.8	2.43	73.0	2.53	46.4
Producer estimated marketing costs[a]	na	na	0.90	27.0	0.52	9.5
Packer-Distributor	0.58	18.1	na	na	na[b]	na
Transport	0.16	5.0	na	na	na	na
Retailer	1.60	50.0	na	na	2.40	44.0
Total retail value	3.20	100	3.33	100	5.45	100

The blueberries in the mainstream and direct market supply chains are conventional. The blueberries in the intermediated supply chain are organic.

[a.] Direct: Includes estimated costs of packing, transportation, and marketing, including labor costs. Total farm per unit revenue is 2.43 + 0.90 = 3.33 ($/lb). Intermediated: Includes packing and transportation costs and estimated opportunity cost of labor for marketing activities. Total farm per unit revenue is 2.53 + 0.52 = 3.05 ($/lb).

[b.] Using a distributor in the intermediated supply chain would add another $0.21 to producer's distribution costs and reduce net farm revenue to $2.32.

Source: King et al. (2010: 24).

the increased availability of blueberries during the local season actually drives down the price.

This is a fresh market product and there is little processing activity associated with the production of the crop. At the farm level, fresh blueberry production remains a labor-intensive system, so there is a substantial wage bill associated with the product from all three supply chains. New Seasons and Hurst's are both located in the local area, and the incomes that accrue to them stay in the local area. Allfoods, in contrast, is not locally owned, so revenue retained by that firm does not stay in the Portland metro area. Because we did not collect sales volume information, it is not possible to compare the total wage and business proprietor incomes associated with the three supply chains.

Food miles and transportation fuel use are lowest in the direct market chain as the average distance from the farm to market is only 10 miles (table 6). The intermediated chain with 70 food miles is lower than the mainstream chain with 115 miles, but fuel use per 100 pounds of product is considerably higher in the intermediated chain because of the smaller loads carried on each trip.

Contributions to social capital in the local area can be most clearly seen in the intermediated and direct market chains. New Seasons, the retailer in the intermediated chain, has a strong community program centered on donating 10 percent of after-tax profits to community groups and activities. The firm has been a leader in a host of local food initiatives. The producer in the intermediated chain, however, expressed the difficulties he faced in becoming deeply engaged in community activities either in the area around his home community or in the Portland area because of time and distance constraints. Thompson Farms has been a cornerstone participant in many local food organizations and has taken on a supporting and mentoring role for immigrant farmers by providing them with land and market outlets. The mainstream firms have some civic engagement but at a more basic level.

Key Lessons

Four general lessons emerge from these case studies of blueberry supply chains in Portland, Oregon:

1. Specializing solely in Northwest-produced and marketed blueberries is insufficient to provide a viable business for any of the participants in these three supply chains. Both the mainstream national retailer and the intermediated retailer treat Northwest blueberries as one component of an annual supply cycle that allows them to sell blueberries throughout the year. The mainstream grower-packer-shipper recognized that his firm needed to provide a year-round supply to his key retail customers and therefore he set up distribution arrangements with firms in other production areas. In a similar

Table 6. Food miles and transportation fuel use in Portland, Oregon—blueberry supply chains

SUPPLY CHAIN SEGMENT	FOOD MILES	TRUCK MILES	RETAIL WEIGHT (CWT)	FUEL USE (GAL)	FUEL USE PER CWT SHIPPED
Mainstream: Allfoods[a]					
Producer	35	70	100	7	0.07
Packer-Distributor	50	100	400	16.7	0.04
Distributor-Store	20	40	400	6.7	0.02
All Segments	115				0.13
Direct: Thompson Farms[b]					
Producer	10	20	15	1.8	0.12
All Segments	10				0.12
Intermediated: New Seasons[c]					
Producer	70	140	18	10.8	0.60
All Segments	70				0.60

[a] Transportation in this chain is in open trucks with a fuel economy of 10 mpg for the segment between the farms and the packing facility and in 48-foot trucks with a fuel economy of 6 mpg for the segments between the packing facility and the distribution center and between the distribution center and the stores.

[b] Transportation in this chain is in a panel truck with a fuel economy of 11 mpg for all trips.

[c] Transportation in this chain is in a delivery van with a fuel economy of 13 mpg for all trips.

Source: King et al. (2010: 25).

fashion, the two locally oriented producers recognize that they must produce a wide range of products in order to extend their marketing period and increase sales even during the berry season.

2. The presence of a strong industry that distributes nationally has a major impact on the functioning of Portland blueberry markets as compared to the other four case study product-place combinations considered in this book. Portland consumers have access to lower prices, more organic product, and more local product sold both through farmers markets and mainstream retailers. Producers, grower-shippers, and retailers benefit from the infrastructure and knowledge that comes with the size of the industry, but they must also accept a high level of competition.

3. The net producer prices are far different in the three supply chains and reflect the need the industry faces, as a whole, to ship most fresh berries out of state. The prices in the mainstream chain are based on global supply and demand conditions. The mainstream producers receive only about 35 percent of the net price the direct market growers receive. Larger mainstream growers could flood local direct markets with lower priced berries but have not done so because they recognize how low their net earnings would be from selling in these limited and labor-intensive markets. The direct market prices reflect local supply and demand conditions, and in looking at those prices, it is important to recognize that the producer estimates his marketing costs are equal to 27 percent of his revenues. The intermediated market prices must also be understood in the context of the competitive pressures of the Portland marketplace. The retailer and producer in this chain ended their long-term relationship when it became clear they could not find a price that was satisfactory for both businesses.

4. In 2009 it was clear that new food safety regulations had the potential to emerge as a significant issue for the direct market and intermediated supply chains. The legislation passed in 2010 took into account the special circumstances of small farms (less than $500,000 in sales) that primarily sell directly to consumers. It remains unclear exactly how this will play out for the two locally oriented farms portrayed in this study. In a somewhat related issue, both the direct

market and intermediated supply chain producers formerly participated in third party certification programs (Food Alliance and organic, respectively) but concluded that these certifications were unnecessary for the local markets that they target.

Notes

1. As discussed later in the chapter, while all five study locations featured small-scale direct-marketed blueberries, Portland was the only MSA located near commercial blueberry production.
2. Hurst's Berry Farm also produces blueberries in Mexico but currently does not export those berries to the United States.
3. By way of comparison the standard discount for quantity purchases of blueberries in farmers markets (factor #2) is 17 percent.
4. This is a pseudonym as the firm prefers to have its identity remain confidential.
5. The total number of firms represented is less than twenty, as some market products under multiple brands.
6. Hurst's grows blueberries on 400 acres in Mexico, but the volume packed in Mexico is much smaller than the Oregon volume because yields are lower and the firm does not pack berries beyond what it produces on its own Mexican acreage.
7. Controlled atmosphere storage for blueberries is only for periods of up to one month, so it is much more limited than controlled atmosphere storage for a product such as apples.
8. The open truck transports only blueberries. The 48-foot trucks may transport other products as well, but the calculations assume that fuel charges are shared proportionally across all products.
9. No reliable estimate is available for the total quantity of blueberries sold through direct markets.
10. The Hurst's berries sold through Allfoods are not regarded as intermediated because the final consumer cannot trace the product back to an individual farm, and the producers do not know that their berries will be sold in the local area.
11. Both are part of what New Seasons defines as the "home grown" region.
12. For the New Seasons deliveries, the vans are generally transporting blueberries alone. If the loads are mixed, the blueberries are being assessed a proportional fuel charge.
13. In 2010 Nature's Fountain established a community supported agriculture program and expanded farmers market sales.
14. Northwest yields are higher than yields in other regions.
15. This assumes that processed berry yields exceed fresh market yields.
16. Direct price comparisons are difficult and may be misleading because some of the berries are organic and others are not.

References

Beaman, J. A., and A. J. Johnson. 2006. "Grocery Retailers in the Northwest." EM 8924, Oregon State University Extension Service. http://ir.library.oregonstate.edu/xmlui /bitstream/handle/1957/20442/em8924.pdf.

Burros, Marian. 2006. "In Oregon, Thinking Local." *New York Times*, January 4. nytimes .com/2006/01/04/dining/04well.html?_r=1&pagewanted=1 (accessed 12/20/2011).

Eleveld, Bart, Bernadine Strik, Karen DeVries, and Wei Yang. 2005. "Blueberry Economics: The Costs of Establishing and Producing Blueberries in the Willamette Valley." EM 8526, Oregon State University Extension Service. arec.oregonstate.edu/oaeb/files /pdf/EM8526.pdf (accessed 12/20/2011).

Kirby, Elizabeth, and David Granatstein. 2009a. "Profile of Organic Crops in Oregon–2008." Washington State University Center for Sustaining Agriculture and Natural Resources. oregonorganiccoalition.org/pdf/OR_OrgCert_Acres_08.pdf (accessed 12/6/2011).

———. 2009b. "Profile of Organic Crops in Washington State–2008." Washington State University Center for Sustaining Agriculture and Natural Resources. csanr.wsu.edu /Organic/WA_CertAcres_08.pdf (accessed 12/6/2011).

King, Robert P., Michael S. Hand, Gigi DiGiacomo, Kate Clancy, Miguel I. Gómez, Shermain D. Hardesty, Larry Lev, and Edward W. McLaughlin. 2010. *Comparing the Structure, Size, and Performance of Local and Mainstream Food Supply Chains*. ERR-99. Washington DC: U.S. Department of Agriculture, Economic Research Service.

Lev, Larry. 2003. "Thompson Farms: Do Real Farmers Sell Direct?" Western Profiles of Innovative Agricultural Marketing. Western Extension Marketing Committee. ag.arizona .edu/arec/wemc/westernprofiles/917%20thompson%20farms.pdf (accessed 12/20/2011).

Lies, Mitch. 2010. "Small Farmers Balk at Food Safety." *Capital Press*, May 13. capitalpress .com/othernews/ml-fda-side-051410-art (accessed 12/20/2011).

Meunier, Andre. 2009 "Blurring the Urban-Rural Line in Damascus." *OregonLive.com*, August 8. oregonlive.com/environment/index.ssf/2009/08/blurring_the_urbanrural _line.html.

Meyers, Brooke. N.d. "Nature's Fountain Farm." http:www.zoominfo.com/s/#!search /profile/person?personId=1439173479&targetid=profile (accessed 9/19/2014).

Northwest Labor Press. 1998. "Oregon Farmworkers Union Signs Historic First Contract." nwlaborpress.org/1998/farmworker.html (accessed 12/20/2011).

Oregon Employment Department. 2011. Oregon Per Capita Personal Income ($), 1998–2012. qualityinfo.org/pubs/single/pcpi.pdf (accessed 2/24/2011).

Pollack, Susan, and Agnes Perez. 2009. *Fruit and Tree Nuts Outlook*. November 24. Washington DC: U.S. Department of Agriculture, Economic Research Service.

Stevenson, Steve. 2009. "Values-Based Food Supply Chains: Red Tomato." agofthemiddle .org/pubs/rtcasestudyfinalrev.pdf.

Terry, Lynne. 2010. "Oregon's Small-Scale Farms Worry about Sweeping Food Safety Bill." *OregonLive.com,* May 30. oregonlive.com/news/index.ssf/2010/05/oregons_small -scale_farms_worr.html.

USDA-ERS. 2009. *Fruit and Tree Nut Yearbook*. Washington DC: U.S. Department of Agriculture, Economic Research Service. ers.usda.gov/publications/FTS/Yearbook09/FTS2009.pdf (accessed 12/20/2011).

USDA-NASS. 2008. "2007–2008 Oregon Agriculture & Fisheries Statistics." Washington DC: U.S. Department of Agriculture, National Agricultural Statistics Service. nass.usda.gov/Statistics_by_State/Oregon/Publications/Annual_Statistical_Bulletin/2008%20Bulletin/stats0708.pdf (accessed 12/20/2011).

———. 2009a. "2008–2009 Oregon Agriculture & Fisheries Statistics." Washington DC: U.S. Department of Agriculture, National Agricultural Statistics Service. nass.usda.gov/Statistics_by_State/Oregon/Publications/Annual_Statistical_Bulletin/2009%20Bulletin/stats0809.pdf (accessed 12/20/2011).

———. 2009b. *Census of Agriculture, 2007*. Washington DC: U.S. Department of Agriculture, National Agricultural Statistics Service.

USDA-NIFA. 2009. "Larry Thompson—Boring, Oregon." csrees.usda.gov/nea/ag_systems/in_focus/sustain_ag_if_profiles_thompson.html (accessed 12/20/2011).

U.S. Government Printing Office. 2009. *Federal Register* 74(142): July 27, 2009. gpo.gov/fdsys/pkg/FR-2009-07-27/pdf/E9-17802.pdf (accessed 12/20/2011).

———. 2011. "Public Law 11-353." gpo.gov/fdsys/pkg/PLAW-111publ353/pdf/PLAW-111publ353.pdf (accessed 12/20/2011).

U.S. Highbush Blueberry Commission. N.d. "Blueprint 2015" (PowerPoint slides). blueberry.org/ushbc/blueprint-2015/Blueprint%202015%20Website%20PowerPoint.pdf?PHPSESSID=a611b75de6b9bf14eb41b9e5c0d0ba28 (accessed 12/20/2011).

Western SARE. 2008. "Thompson Receives Madden Award." *Simply Sustainable* 2 (March): 1, 6–7. westernsare.org/Learning-Center/Newsletters/Western-SARE-Newsletter-Archives/Spring-2008-Simply-Sustainable (accessed 12/20/2011).

5 Spring Mix Case Studies in the Sacramento MSA

Shermain D. Hardesty

Introduction

The following case studies describe three supply chains for spring mix in the Sacramento Metropolitan Statistical Area:

an upscale regional supermarket chain (mainstream supply chain),
a local producer selling at a farmers market (direct market supply chain), and
a natural foods grocery cooperative selling locally grown spring mix (intermediated supply chain).

For this study, "local" refers to spring mix grown, processed (if necessary), and shipped by firms within the Sacramento area; this definition distinguishes it from spring mix grown in the Salinas Valley, which is approximately 175 miles (a three-hour drive) from downtown Sacramento. Most of the local spring mix is grown in western Yolo County, primarily in the Capay Valley.

The Location: Sacramento Area

The Sacramento area is made up of four counties—Sacramento, Placer, El Dorado, and Yolo—located in the northern Central Valley (see figure 11). It is the 26th largest metropolitan area in the United States, with a population of 2,109,832 in 2008 (U.S. Census Bureau 2010). Most of the Sacramento area is part of the Sacramento Valley, which is the northern portion of the long and flat Central Valley—home to much of California's most productive agriculture.

Local region

Primary spring, summer and fall growing regions

Primary winter growing regions

CALIFORNIA

ARIZONA

Fig. 11. Western U.S. production regions for spring mix.

While Sacramento is well-known as a center for government services within California, the Sacramento area has a diverse base of employers and industries. The city of Sacramento is at the hub of a major highway corridor; Interstate 5 runs north-south through the city, while Interstate 80 stretches east to west. Many food companies maintain distribution warehouses in the area. Farmer's Rice Cooperative, Campbell's Soup, Blue Diamond, Crystal Creamery, and Pacific Coast Producers are the area's major food processors.

Raley's, the nation's 38th largest grocery chain, is headquartered in the region. Seven other top 75 North American food retailers market in the Sacramento area, including another regional supermarket company, Savemart, and national companies Wal-Mart, Costco, Safeway, Whole Foods, and Trader Joe's (Supermarket News n.d.). Specialty grocers include

Nugget Markets, three natural foods cooperatives, and numerous small ethnic markets.

According to the USDA's 2007 Census of Agriculture (USDA-NASS 2007), agricultural production within the four-county region totaled $795 million on 5,152 farms, of which 4,726 (91.7 percent) were classified by the USDA as "small," meaning with annual revenues under $250,000. Major food commodities produced in the Sacramento area are processing tomatoes, wine grapes, rice, milk, almonds, apples, and pears.

Interest in locally grown food is strong in the Sacramento area. There are Slow Food chapters in Yolo, Sacramento, and Placer Counties and at Lake Tahoe. Regional agricultural promotional programs include Capay Valley Grown, Apple Hill Growers Association, and Placer Grown. The four counties have 36 farmers markets (USDA-AMS n.d.), several of which operate year-round, and about 30 community supported agriculture (CSA) programs (Local Harvest n.d.). In 2007, 14 percent of the farms in the Sacramento area were involved with direct marketing, compared to 9 percent nationally. The region's direct marketers averaged $19,518 in revenues from this channel, but they ranged widely by county, from a low of $6,924 in Placer County to a high of $66,568 in Yolo County (USDA-NASS 2007).

The Product: Spring Mix

Spring mix was chosen as the product focus for the supply chain case studies for the Sacramento area, rather than the general category of leafy greens. Spring mix originated in France's Provence region, where it was called *mesclun*. Alice Waters, founder of the Berkeley restaurant Chez Panisse, is often credited with introducing spring mix on the West Coast.

There is no standard of identity for spring mix. A leading marketer, Earthbound Farm, lists the following organic greens as ingredients for its spring mix, with the caveat that ingredients in each package may vary: baby lettuces (red and green romaine, red and green oak leaf, lollo rosa, tango), red and green chard, mizuna, arugula, frisée, and radicchio. The proposed National Leafy Greens Marketing Agreement indicates that spring mix consists of "mixtures of baby lettuces, mustards, chards, spinach, and chicories that vary based on availabilities" (USDA-AMS 2011a: 4).

Spring Mix Production

Spring mix is grown primarily in California, Arizona, and Florida. The largest production region for spring mix is California's Salinas Valley (Monterey and San Benito Counties). Spring mix production regions in the western United States are identified in figure 11, along with the Sacramento area. Spring mix is planted from January through October in the Salinas Valley and the Santa Maria area and harvested from February through November. In Fresno County, Imperial County, and Arizona's Yuma County, spring mix is planted from October through February and harvested from November through March.

Spring mix production data for 2008 in the major California counties are shown in table 7. No spring mix production data are reported for any other locations in the United States. Monterey County in California's Salinas Valley is the largest producer of spring mix in the nation (and probably in the world); in 2008, the county produced 242 million pounds of spring mix on 12,900 acres, with a farm gate value of $172 million. Due to double cropping, yield per acre in Monterey County averaged 18,800 pounds in 2008, compared to 7,000 pounds in Imperial County. Although spring mix is considered a niche crop, grower revenues in 2008 exceeded those earned nationwide for more traditional vegetable crops, such as green peas and asparagus.

Spring Mix Processing

After being harvested, baby leafy greens are trucked to a nearby processing facility where they are washed, dried, mixed together, packaged, and shipped to customers within 48 hours in refrigerated trucks.[1] Some grocery chains and food service operations buy the spring mix directly from processors, while others purchase the product from distributors who bought it from processors or producer wholesalers. With proper handling, packaged spring mix has a 17-day shelf life.

While most of the spring mix grown in California is machine harvested, and machine processed and packaged at large processing facilities in the Salinas Valley, small farmers throughout the state produce an unspecified amount of spring mix by hand-harvesting and mixing a variety of baby lettuces with little or no processing equipment, and marketing it in bulk form—often at farmers markets and/or through CSA programs.

Table 7. Estimated spring mix production in major California counties, 2008

COUNTY	ACRES	POUNDS	FARM GATE VALUE($)
Monterey	12,901	242,000,000	172,386,000
San Benito	2,591	49,178,000	16,419,000
Imperial	6,306	44,142,000	37,521,000
Total	21,798	335,320,000	226,326,000

Source: Monterey County Agricultural Commissioner, 2008 Crop Report. San Benito County Agricultural Commissioner, 2008 Crop Report. Imperial County Agricultural Commissioner, 2008 Crop Report.

Spring Mix Consumption

No per capita consumption data are reported for spring mix. However, the combined production of spring mix in California's Monterey, San Benito, and Imperial Counties of 335.3 million pounds in 2008 converts to 1.1 pounds of annual per capita consumption for the United States. Comparison of production levels in 1998 and 2008 in two of the counties, Monterey and Imperial, indicates a 202 percent increase. It is likely that growth in consumption nationwide mirrors the tripling in production over the nine-year period.

Mainstream Case: Nugget Markets

Nugget Markets Inc. is a regional chain owned by the Stille family, who founded the firm in 1926. It operates nine upscale Nugget supermarkets and three Food4Less warehouse-type discount stores; eight Nugget stores and one Food4Less store are in the Sacramento area. Nugget's sales revenues totaled $288 million in 2009. In the Sacramento-Stockton-Modesto grocery market, it is ranked number six with a 2.9 percent share; locally owned Raley's is the market leader with a 23.4 percent market share, followed by Safeway with 19.6 percent and Savemart with 17.3 percent.

Nugget describes itself on its web site (nuggetmarket.com) as "a family of dedicated people with a love of food and a passion for excellent service." In the late 1990s Nugget overhauled its marketing strategy with a new "Fresh to Market" concept and began using an open European-style store format featuring specialty departments and a large selection of both conventional and organic fruits and vegetables displayed attractively in its produce department. Locally grown produce is marked with colorful signage.

Supply Chain Structure, Size, and Performance

Earthbound (ebfarm.com) is Nugget's primary spring mix brand. Nugget's spring mix supply chain is depicted in figure 12.[2] This case study focuses on production, processing, marketing, and distribution for this supply chain.

Production and Processing

Earthbound is located in the Salinas Valley, which is often called "America's salad bowl." Now an icon in the organic food industry, Earthbound was founded in 1984 by Drew and Myra Goodman as a two-and-a-half-acre Carmel Valley garden plot. In 1986 Earthbound became the first company in the nation to sell small bags of prewashed mixed organic baby lettuces to retail customers, starting with natural foods stores in the Monterey Bay area as well as in San Francisco. In 1992 Earthbound purchased a 32-acre farm and built a 9,000-square-foot processing facility. Bagged mixed salad greens began growing in popularity nationwide, and by 1993 Earthbound was selling one-pound bags of salad greens to Costco, Safeway, and Albertsons. In 1999 the Goodmans became equal partners with produce giants Mission Ranches and Tanimura & Antle, and operated Earthbound as Natural Selection Foods.[3] In July 2009 HM Capital Partners LLC, a private equity firm focused on investments in the food, energy, and media industries, acquired Natural Selection Foods and changed the firm's name to Earthbound Farm (Withers 2009).

All of Earthbound's produce is now certified organic. While Earthbound describes itself on its web site as the world's largest grower of organic produce, it actually sources its leafy greens and other produce from 150 farms on more than 37,000 acres (Earthbound Farm 2011: 4). It has become a diversified food company, marketing more than 100 different varieties of organic salads, fruits, and vegetables and a small assortment

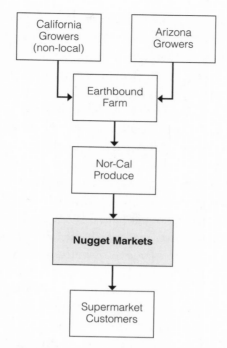

Fig. 12. Mainstream supply chain, Nugget Markets.

of dried fruits, snacks, cookies, and granola. Its broad product offerings reflect the company's mission: "to bring the benefits of organic food to as many people as possible and serve as a catalyst for positive change" (Earthbound Farm 2011: 4).

The front of Earthbound's spring mix packages includes the USDA organic logo and a label indicating that the product was "grown in the USA and Mexico and processed in the USA." Approximately 60 percent of Earthbound's spring mix is grown in the Salinas Valley, located about 175 miles from downtown Sacramento. During the late fall and winter, spring mix is produced in the "desert region," which consists of Imperial County in southeastern California, neighboring Yuma County in Arizona, and northern Mexico.

Substantial coordination is required in the planting and harvesting of spring mix crops to ensure that the quantity and variety of the salad greens harvested match up with its expected demand. Food safety considerations and the diminishing supply of field labor have caused more growers to utilize stainless steel harvesters. The delicate baby leafy greens are machine-harvested early in the morning when temperatures are the coolest. The cut greens move through the machine's rollers and onto a two-stage sort belt, which drops out undersized and partially cut leaves; a fine water mist is applied before the leaves fall into totes. According to Valley Fabrication, a major manufacturer of such harvesters, the machine can harvest about 6,000 pounds per hour using totes (Valley Fabrication n.d.). Another manufacturer of spring mix harvesters, Ramsay Highlander, claims that its machines are capable of harvesting over 15,000 pounds per hour, and that spring mix harvesting costs have been reduced from $0.28 per pound to less than $0.01 per pound (Ramsay Highlander n.d.). The totes are packed onto pallets and loaded onto trailers. The harvested leafy greens are transported to a large refrigerated box next to Earthbound's processing facility (see figure 13).

Assuming that 60 percent of the spring mix crop was grown in Monterey County and 40 percent in Imperial County, growers received farm gate prices averaging $0.77/lb in 2008.[4] Assuming that organic product earned a 10 percent premium and made up 45 percent of each county's reported production, the estimated average price paid to growers for organic spring mix was $0.81/lb.

Processing

Because Earthbound sources leafy greens from two growing regions, it operates two processing facilities, one in San Juan Bautista, California (at its headquarters), and a smaller one in Yuma, Arizona. Staff usually move from its processing facility in San Juan Bautista to its Yuma plant (about 585 miles) in mid-November to process leafy greens grown in the desert region, and usually move back to San Juan Bautista sometime during mid-March.

After being cooled, the leafy greens are delivered to Earthbound's processing facility (see figure 13). As described in Earthbound's Media Kit (Earthbound Farm 2011: 8–9), samples of the greens are taken from each load for testing for food-borne pathogens. The greens are inspected, washed in chlorinated water, and dried on Earthbound's custom-designed equipment line. Samples of the packaged salads are removed for testing

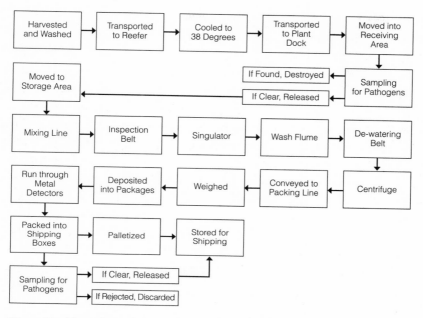

Fig. 13. Earthbound Farm's processing flow. Source: Adapted from Seltzer, Rush, and Kinsey (2009: 9); used with permission of the Food Industry Center.

for food-borne pathogens. All packaged product is held until clear test results are obtained. Individual packages are packed in cases, which are then palletized. The pallets are moved to a large refrigerated storage area next to the truck loading docks. Large supermarket chains pick up orders from Earthbound and deliver the loads to their produce distribution centers. Produce distributors also purchase from Earthbound, as do produce wholesalers located at terminal markets. Trucking to the East Coast takes three and a half to six days.

Distribution

As noted on its web site (nuggetmarket.com/products/produce/), Nugget has "been working with Nor-Cal for more than two decades on a relentless search for the best produce grown locally and across the globe." Nor-Cal is a West Sacramento–based business that was founded in 1971 by Daniel and Linda Achondo. Their sons, Dan and Todd, are now actively involved in Nor-Cal's management. The firm began as a vegetable distributor, added fruit in 1990, then added organic produce in 2000. In 1999 Nor-Cal expanded its geographic reach to include Brookings, Oregon, to the north; Salinas in the south; Minden, Nevada, to the east; and San Francisco to the west. It built its current 85,000-square-foot facility in 2002, which includes a 5,000-square-foot office. In 2008 Nor-Cal grossed $121 million. Nugget is Nor-Cal's largest revenue source.

Nor-Cal has three cold rooms. Naked lettuce is stored in a cold box (wet room) held at 34.5 degrees. Packaged salads are held in a different room at 34.5 degrees. Alarms go off in these rooms if the temperature varies by more than 1.5 degrees.

As a general practice, in order to minimize its quality control problems Nor-Cal uses a relatively small number of suppliers who know its quality expectations. Nor-Cal reviews its produce inventory daily. Its buyers arrive at work between 4:00 a.m. and 6:30 a.m. and begin receiving orders from stores. They phone in most of their orders to their suppliers, including Earthbound, by 9:00 a.m. Nor-Cal's order entry staff enters each store's orders. The dispatcher contacts its Salinas-based trucking company. Trucks get load sheets and begin arriving at coolers in Salinas around 1:00 p.m.; they often go to more than one supplier. The drivers check their load sheets to make sure that proper items are loaded onto the truck; they do not do a quality check. Nor-Cal takes ownership of the product when this loading is done. The trucks from Salinas arrive at Nor-Cal by 11:00 p.m. During the desert season the trucking company moves part of its trucking fleet to El Centro; the trucks begin picking up by 1:00 p.m. from various Nor-Cal vendors, including Earthbound's Yuma plant, and arrive at Nor-Cal's warehouse the next afternoon.

Since temperature control is key for many produce items, Nor-Cal's receiving personnel use their hands to feel the temperature of truck trailer when it arrives; if it feels too warm, they check the current temperature inside the refrigerator unit and inside a box of produce. The pallets are unloaded into the warehouse using a pallet jack and forklift; boxes are spot-checked for quality.

Around 11:00 p.m. the same evening, warehouse forklift operators begin loading Nor-Cal trucks using pick lists based on order entry sheets. Nor-Cal truck drivers leave the warehouse with invoices; the last load goes out around 4:00 a.m. When the truck arrives at a Nugget store, the driver unloads and has store personnel sign the invoice. Nor-Cal transfers product ownership to Nugget and other customers 24 hours after it has delivered its orders. On the rare occasion that Nor-Cal has quality problems with products from Earthbound, it contacts its Earthbound sales representative to discuss the problem and provide the lot code information. Nor-Cal requires considerable operating capital; for its purchases, the terms are net 10 days, while the terms for its sales are net 14 days.

Marketing and Retail Operations

Earthbound has succeeded in gaining distribution for its products in over 75 percent of U.S. grocery stores. It also has a growing private label program (Withers 2010). Nugget stores display bulk Earthbound spring mix in a large bowl (labeled only as "spring mix," with a USA country of origin designation) alongside other organic produce. In a large refrigerated unit Nugget displays five-ounce and one-pound clamshells of Earthbound spring mix, along with a variety of other Earthbound packaged salads, five-ounce clamshells of another organic spring mix brand, five-ounce bags of another brand of conventional spring mix, and other packaged salads and numerous pre-cut vegetable and fruit products. Earthbound's five-ounce product is labeled as "mixed salad greens," although it is identical to the one-pound product, which is labeled "spring mix." The target temperature for the refrigerated case and the backroom walk-ins at the Nugget stores is 41 to 42 degrees. Although Nugget does market some locally grown produce, such as apples, mandarin oranges, and heirloom tomatoes, it does not sell any locally grown spring mix.

Nugget's only purchasing contract related to spring mix is with the firm that processes its conventional packaged salad products (including spring mix); the contract provides fixed prices on products and includes quarterly rebates to Nugget based on its purchasing volume. All the products are supplied to Nugget by Nor-Cal with fixed prices. The only recent price adjustment occurred in 2008 when fuel costs soared. Nor-Cal receives promotional ad discounts each week from Earthbound that cycle between the different products: $1.00 for bulk spring mix (three-pound case), $1.20 per case of eight five-ounce packages, and $3.00 per case of six one-pound packages. Projected sales volumes at all Nugget stores (based on data provided through November 20, 2009) of Earthbound spring mix products are displayed in table 8. With projected sales of almost $400,000, spring mix will make up 0.14 percent of Nugget's total revenues.

Thus Nugget's merchandising is generating significant margins from Earthbound's spring mix products. The bulk product is identical to the spring mix in clamshell packages and has the same retail price as the one-pound clamshell. It represents 59 percent of Nugget's Earthbound spring mix sales and 63 percent of the weight volume; it has the highest gross margin, although it lacks the packaging protection and refrigeration of the clamshell product. The five-ounce product contributes only 12 percent to spring mix revenues and 7 percent of the weight volume. Consumers apparently perceive the spring mix displayed loose in bowls as fresher than the packaged product and/or prefer to control the amount of spring mix that they purchase. Additionally, the old adage of "buy big to save" holds with regard to spring mix; the per-pound price of the spring mix in the five-ounce clamshell is almost double that of the one-pound clamshell.

Based on an average wholesale price of $2.74 per pound for its bulk spring mix, Nugget is capturing the largest share of the retail price $3.75 (57.8 percent). Earthbound's share is estimated to be $1.16 (17.9 percent), followed by the producers' share of $0.81 (12.5 percent), which includes freight costs ($0.02). In this mainstream supply chain, the distributor Nor-Cal earns the smallest share of the retail price—$0.77 (11.9 percent), which includes freight costs of approximately $0.50.

Food Miles and Transportation Fuel Use

The Earthbound spring mix sold by Nugget typically moves through the four-segment supply chain from the grower to Earthbound and then to Nor-Cal before reaching the retail location (Figure 12). Based on an average trip of 30 miles from the Salinas Valley grower's field to Earthbound's processing facility with 130 hundredweight (cwt) at 6 mpg fuel efficiency, the food miles total 30 miles and fuel usage is 0.08 gal/cwt; the comparable 45-mile trip in the desert requires 0.12 gal/cwt. Food miles from San Juan Bautista to Nor-Cal's warehouse in West Sacramento total 192 miles; the tractor-trailer rig hauling 400 cwt achieves 5.5 mpg using 0.17 gallons/cwt for the 372 truck miles. The comparable segment from

Table 8. Nugget's Earthbound spring mix sales, 2009

PRODUCT	SALES (LB)[a]	PAID TO NOR-CAL ($)	RETAIL ($/LB)	% OF TOTAL EARTH-BOUND (LB)	% OF TOTAL EARTHBOUND ($)
5 oz clamshell	3,741	24,030	12.77	6.6%	12.1%
1 lb clamshell	17,600	64,855	6.49	30.9%	29.0%
Bulk	35,646	97,651	6.49	62.6%	58.8%
All Earthbound spring mix	56,986	186,536	6.90		

[a.] Projected, based on sales through November 20, 2009.

Source: Case study interviews.

the desert region involves 618 food miles and 1,250 truck miles, requiring 0.57 gal/cwt. The journey of 16 food miles from the warehouse to the main Nugget store in Davis in a 250 cwt tractor-trailer rig with 6 mpg fuel efficiency requires 0.10 gal/cwt, traveling 150 truck miles overall. Total fuel usage per cwt is 0.35 and 0.79 gal/cwt, respectively, for Salinas Valley–grown and desert-grown Earthbound spring mix. Large loads for each segment keep total fuel use per cwt low.

Community and Economic Linkages

As a family-owned local firm, Nugget provides significant support to the local community. Its annual donations total approximately $450,000 in cash and in-kind contributions, far outstripping the retail giant Wal-Mart, which reported donations to local organizations totaling $41,500 (*Davis Enterprise* 2010). Nugget's primary activity related to local foods is being a top sponsor of the annual Village Feast, which is organized by Slow Food Yolo and Davis Farm to School (as its major fundraiser), with food sourced from Davis Farmers Market vendors. Nugget also supports various agricultural organizations, including the Center for Land-based Learning and the Good Life Garden at University of California–Davis, as well as food banks, homeless shelters, soccer programs, Little Leagues, high school graduation night celebrations, and various other community organizations. Nor-Cal's primary support activity is donating surplus and/or "expired" produce to the local Senior Gleaners.

As local firms, Nugget and Nor-Cal have significant local economic impacts, although the impacts are clearly attributable to a wide range of products, not just spring mix. Nugget employs approximately 1,500 people, 60 percent of whom are full-time; its payroll totals approximately $40 million. It is recognized as a company that empowers its employees; Nugget has been rated among the top 100 firms to work for by Fortune for the past five years, earning its highest ranking of fifth in 2010 (CNN Money n.d.). It pays approximately $1.4 million in property taxes. Nugget's chief financial officer estimated that 3 to 4 percent of Nugget's purchased goods are produced locally. Nor-Cal has 140 employees, with a payroll of approximately $6 million. It pays approximately $140,000 in property taxes to Yolo County. Its major purchase from a local firm is about 340,000 gallons of diesel fuel annually for its 29 delivery trucks.

Prospects for Expansion

Nugget grew substantially prior to the recent recession. While it is likely that Nugget will expand its offerings of locally grown produce and source most, if not all, of this product through Nor-Cal, it appears unlikely that there is (or will be in the near future) sufficient locally grown spring mix available to supply Nugget's stores; this situation is discussed further in the intermediated spring mix supply chain case.

Nugget, Nor-Cal, and Earthbound are impacted by regulatory pressures. Earthbound has become very vigilant with its food safety practices because it packed the spinach that was determined to be the source of the highly publicized and tragic *E. coli* O:157 outbreak in 2006. It is a signatory of the California Leafy Greens Marketing Agreement, which is serving as the model used by USDA for a National Leafy Greens Marketing Agreement. The California program is also being considered as a model by the Food and Drug Administration as it begins to develop national safety standards for the growing, harvesting, and packing of fresh produce.

Nugget's produce director indicated that Nugget is very proactive with respect to food safety issues. There are both time and paperwork costs to food safety regulations, but Nugget expects its standards to be superior to regulatory standards. He mentioned that many local county environmental health departments are inconsistent in their enforcement of their rules.

Nor-Cal is concerned about the cost implications of regulations. Its management stated that the recently implemented standards by California's Air Resources Board cost Nor-Cal $120,000 to retrofit its trucks.

Key Lessons

Two general lessons emerge from this case study of the mainstream supply chain for spring mix. Over the past 25 years Earthbound has built the nation's spring mix market from a tiny start-up to having distribution in 75 percent of grocery outlets nationally. The company is a highly reliable year-round supplier of organically grown leafy greens that are susceptible to a variety of weather and pest problems. Earthbound coordinates the crop flow from numerous spring mix growers in two different production regions and annually executes the complex transfer of its processing operations and many of its management and technical employees from the Salinas Valley to the desert region. As in many other produce companies, the employees have to endure the hardships associated with splitting their residences in two areas. Earthbound has met its mission to "bring the benefits of organic food to as many people as possible."

Second, Nor-Cal and Nugget have created a strong and highly cost-effective relationship as Nugget has simplified its produce distribution operations by partnering with Nor-Cal for the past 20 years. It relies on Nor-Cal to manage the purchase and transport of a highly diverse array of high quality produce. Nor-Cal's large truckloads minimize the hauling and handling costs. Each of Nugget's produce managers communicates directly with a contact at Nor-Cal. The consistency of Earthbound's supplies alleviates the need for much dialogue between Nor-Cal and Earthbound.

Direct Market Case: Fiddler's Green Farm

Fiddler's Green Farm is a small organic farm located in Yolo County's Capay Valley, approximately 60 miles from downtown Sacramento. In 2010 there were at least 28 farming and ranching operations in Capay Valley; most of these producers farm organically and are involved in direct marketing.

Cliff and Marian Cain founded Fiddler's in 1978. In 1982 the farm became the first Capay Valley farm to be certified organic. The Cains retired in 1989 and sold the farm to Steve and Sue Temple. After earning a BA in philosophy at UC Davis, Jim Eldon joined Fiddler's in 1991 as the farm manager. He quickly became a 50 percent owner in Fiddler's and immediately began selling at the Davis and Marin Farmers Markets. In 1996 Eldon and his wife, Julie Rose, became the sole owners of the 37-acre farm.

After acquiring sole ownership of Fiddler's Green, Eldon expanded the operations by leasing an additional 25 acres nearby. He sold some produce wholesale but mainly direct-marketed his produce through farmers markets and a CSA program that he launched in 1992. During the spring of 1999 Eldon lost all his crops due to a deep freeze and was forced to discontinue his CSA and lay off all but one of his 15 employees.

Currently Eldon is farming on 32 acres; he and Julie live on the farm with their two children. Their son has been helping out since he was seven years old; now 16 years old, he often works at the farmers markets with Eldon. Fiddler's has four full-time employees who work about eleven months of the year; they do not have any health insurance. The longest-term employee has housing on the farm. In 2008 Fiddler's grossed about $120,000, which is substantially less than the $500,000 it generated before the 1999 crop disaster.

Fiddler's is of particular interest as a local foods case study because it is a direct marketer like many small farms. Furthermore, Eldon believes that it is too small to be profitable and is seeking ways to increase his revenues and profitability without farming any additional acreage.

Supply Chain Structure, Size, and Performance

Product flows within Fiddler's supply chain are shown in figure 14. Fiddler's plays the central role in the chain, determining what to produce, growing all of the produce, setting retail prices, conducting all of the direct marketing, and maintaining relationships with all of its customers.

Production

Fiddler's produces 90 to 100 different crops annually, including a variety of lettuces, leeks, carrots, asparagus, melons, summer squash, peas,

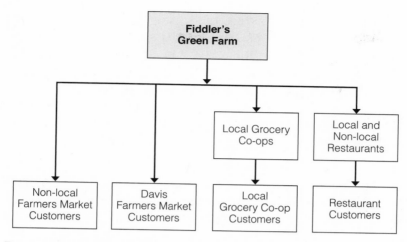

Fig. 14. Direct market supply chain, Fiddler's Green Farm.

numerous salad greens, and several types of bok choy. Eldon grew 45 crops during the fall of 2009. He remarked that one of the hardest things he had to learn when he started farming by himself was what to grow and when to plant.

Fiddler's major equipment items include a 55-horsepower tractor, small Caterpillar tractor, seeder, and mower. The buildings include a pole barn, packing shed, two coolers, and an old tool shed. Eldon is usually planting crops nine months of the year. He uses a cold frame to start crops in the late winter. Eldon plants spring mix in 300-foot beds during the spring and expands to 600-foot beds during the summer.

Harvesting and Packing

To prepare for a farmers market Eldon puts together a load list of crops to take to the market. Fiddler's employees harvest all the crops the day before except asparagus and peas, which can be stored in the cooler for several days. As they are harvested, all leafy greens crops (such as the lettuces for spring mix) are covered immediately with wet burlap to retain their moisture. All the harvested crops are brought into the packing shed, hand-dunked in a 500-gallon stock tank, and then rinsed. Employees bundle crops such as beets, bok choy, and carrots into bunches. They pack produce for the grocery cooperatives and restaurants into standard pack boxes. Greens for the spring mix are drip-dried and loosely packed in four-pound boxes; the boxes are stacked tipped up to drain out any residual water. Eldon takes pride in being recognized for having produce that is "backyard garden fresh and clean." When the weather is cool, Eldon loads his truck the evening before a farmers market.

Marketing and Distribution

Fiddler's markets its produce through three channels: farmers markets, grocery cooperatives, and restaurants. Eldon sells at three farmers markets: the Davis Farmers Market on Saturdays and the Marin Farmers Markets (not local) on Thursdays and Sundays. In 2008, revenues totaled approximately $45,000 at the Davis Farmers Market and $20,000 at each of the Marin Farmers Markets. Sales to two local grocery cooperatives and restaurants (one of which is local) generated the remaining 30 percent of the revenues.

According to USDA's 2007 Census of Agriculture, producers in California led the nation in revenues generated from sales directly to consumers (USDA-NASS 2007). While the direct marketers in Yolo County averaged $66,568, Fiddler's exceeded this figure with direct marketing revenues totaling approximately $85,000.

The Davis Farmers Market (davisfarmersmarket.org/) was established in 1976. The market has a permanent covered area where most of the farmers have stalls. It has a waiting list of new producers, and market guidelines provide stall space priority to local farmers, while maintaining a diverse mix of products. It is not uncommon for the market to have 7,000 to 10,000 visitors during a single week. During the peak of the summer, there are about 45 to 50 farmers selling at the Saturday market, compared to about 40 farmers during the fall and spring and about 30 farmers during the winter. The Davis Farmers Market was voted the nation's favorite large farmers market in 2009 in a contest organized by the American Farmland Trust.

The Davis Farmers Market is operated by the Davis Farmers Market Association, which runs the Saturday Market, Wednesday "Picnic in the Park" (from mid-March through October), Wednesday Winter Market (from November through mid-March), and Wednesday Market at the UC Davis Quad during fall and spring quarters. Members pay a $25 annual fee to the association. The association charges its farmer members a stall fee equaling 6 percent of their sales for the Saturday market, with a $30 per day minimum; nonmembers cannot sell at the Saturday market.

As certified farmers markets, both the Davis and Marin farmers markets require that farmers can only sell what they grow. Farmers must obtain their certificates annually from their county agricultural commissioner and submit the certificate to the market manager. The scale used at the farmer's stand must bear a current seal from the county sealer of weights and measures. A family member or employee may sell for an approved seller, and employers or employees may be required to provide proof of employee status to the Davis Farmers Market manager. The Davis Farmers Market also requires that sellers display a sign bearing the producer's business/farm name and county of origin and that they carry $1 million of liability insurance. Sellers who sample their products must abide by the regulations established by the county's Environmental Health Department. The market also requires that sellers of "salad mix" post a notice indicating that the mix is field harvested and should be rinsed before serving; this notice is not required if the salad mix has been washed in a certified kitchen.

On Saturdays Eldon leaves the farm at 5:15 a.m. and arrives at the Davis Food Co-op by 6:15 a.m. After unloading several boxes of produce, he drives less than half a mile to the Davis Farmers Market, which is located at the city's Central Park. The market operates year-round, and Eldon tries to maintain a year-round presence at this local market. Eldon is one of about a dozen farmers at the Saturday Davis Farmers Market who sell primarily vegetables. Five of the farms sell spring mix during various times of the year; three of them are organic. Prices for the organic spring mix range from $5.00 to $8.00 per pound. Eldon usually brings 40 pounds of spring mix to the Davis market and sells it for $8.00 per pound; between mid-June and mid-October in 2009, he was the only vendor at the market selling spring mix. Fiddler's has a unique microclimate that enables Eldon to produce spring mix during the summer when it is too hot for other local farms to do so.

The Saturday market closes at 1:00 p.m. After taking down his stall, loading up his truck, and turning in his load list, Eldon drives less than a quarter of a mile to make a delivery at a downtown Davis restaurant,

Table 9. Fiddler's Green Farm marketing costs and revenues, 2008

MARKETING DAY & ACTIVITIES	HOURS OR MILES	RATE ($)	WEEKS
Saturday—Davis Farmers Market, Davis & Sac Food Co-op, Restaurant			
Eldon's labor	11	18.83	46
Packing shed workers	8	9	46
Transportation	95	0.637	46
Market fees			
All Saturday marketing activities			
Both Sunday and Thursday—Marin Farmers Market, Restaurants			
Eldon's labor	20	18.83	36
Packing shed workers	16	9	36
Transportation	350	0.637	36
Market fees			
All Marin Farmers Market marketing activities			
Total Marketing Activities			

Source: Case study interviews.

Tucos Wine Bar and Café. Then he heads to Sacramento to make a delivery at the Sacramento Natural Foods Co-op before returning to Fiddler's around 4:15 p.m.

On Thursdays and Sundays Eldon leaves Fiddler's at 5:15 a.m. and drives 175 miles roundtrip (about three hours of driving time) to the Marin Farmers Market at the Civic Center in San Rafael (marinfarmers markets.org). He often picks up a friend in Point Reyes to help him at the busy market. The market operates from 8:00 a.m. to 1:00 p.m. His restaurant customers pick up their produce at the market's designated parking lot for chefs. Marin's Sunday market is the third largest farmers market in California. During its peak summer season, nearly 200 local farmers, specialty food purveyors, and artisans are selling at the market.

2008 COST ($)	REVENUES ($)	MARKETING COSTS/ REVENUE %
9,528		
3,312		
2,784		
2,725		
18,349	70,000	26.2%
13,558		
5,184		
8,026		
2,400		
29,168	50,000	58.3%
47,517	120,000	39.6%

Eldon remarked that he sold 600 pounds of melons at this market on a busy summer Sunday in 2009.

Usually Eldon brings only 10 pounds of spring mix to the Marin market and sells it for $6.00 per pound. He competes with growers from Sonoma County and the coastal South Bay region who have more favorable growing conditions. The Marin Farmers Markets operate year-round, and Eldon normally sells at these markets from mid-March through December.

Eldon commented that he enjoys talking to new and returning customers at the markets. New customers often inquire about the size and location of his farm and his farming history. Both new and returning customers often ask questions about specific produce items, especially about how they taste and ways to prepare them. Numerous customers have asked him to start a CSA. It was at these farmers markets that Eldon made connections to the restaurant chefs to whom he sells.

Since Fiddler's coordinates its marketing activities in different channels during three marketing days, it is helpful to separate the marketing costs and revenues for each day (table 9). A typical busy Saturday for Eldon requires approximately 95 miles of driving roundtrip and 11 hours of Eldon's time for driving, setting up, selling, and taking down at the Farmers Market and delivering produce to a restaurant and two local food cooperatives. His workers spend eight hours preparing the loads and load the truck; the average cost of this labor is $9.00 per hour. Setting the cost of vehicle operation at $0.637 per mile and the opportunity cost of Eldon's labor at $18.83 per hour, the total cost for 46 trips over the course of a year is $18,349 (including stall fees at the farmers market). With annual sales through the channels served by these deliveries totaling $70,000, this cost represents 26 percent of Fiddler's associated sales revenues.

Eldon's Thursday and Sunday marketing days require 10 hours of Eldon's time for driving, setting up, selling, and taking down at the Farmers Market. All other marketing costs are the same as those on Saturday, except that his roundtrip mileage is 175 miles. The total cost for 36 trips over the course of a year is $14,584 for each market day (including stall fees at the farmers market). With annual sales through the channels served by these deliveries totaling $25,000, the marketing costs represent 58 percent of Fiddler's associated sales revenues.

Sales to the grocery co-ops on Saturdays bolster Fiddler's revenues and bring down its marketing expenses relative to revenues. Recent research of other organic direct marketers in the area indicates that Fiddler's 40 percent marketing expenses-to-revenues ratio is not unusually high, and that its comparable ratio on Saturdays of 26 percent demonstrates significant marketing effectiveness (Hardesty and Leff 2010).

Fiddler's produces about 2,000 pounds of spring mix, which accounts for about 11.5 percent of the farm's total revenues. Fiddler's allocation of revenues is shown in table 10 for spring mix by market channel.

Based on his recent sales of spring mix to the Sacramento Natural Food Cooperative, Eldon figured that he could sell 240 pounds per week to the store for retail sales and its deli. It appears, however, that selling to this cooperative for $4.00/lb is not optimal since his net revenues from

Table 10. Fiddler's spring mix revenues by channel

CHANNEL	$/LB	ESTIMATED REVENUES ($)
Davis Farmers Market	8	8,800
Marin Farmers Markets	6	3,600
Restaurants	5	1,000
Grocery Co-ops, Davis & Sacramento	4	400
All Channels		13,800

Source: Case study interviews.

selling the spring mix at the Davis Farmers Market for $8.00/lb, after subtracting 26.2 percent for his marketing costs, are $5.90/lb.

Like other small farms in the area, Fiddler's does not segregate production costs for the wide variety of crops that it grows. No cost studies have been prepared for such diverse farming operations or for spring mix. Eldon is considering how to become profitable without expanding his acreage. His concern is valid considering the 2009 payroll expenses and revenues estimated at $90,000 and $100,000 respectively. The $10,000 balance does not cover Fiddler's estimated marketing expenses (which total more than $29,000 when adjusted for the packing shed labor already counted as payroll expenses). It also does not cover nonlabor production input costs and various overhead expenses. Fiddler's revenue shortfall in 2009 due to the unavailability of tomato and pepper transplants in the spring contributed to this difficulty.

Food Miles and Transportation Fuel Use

Eldon integrates deliveries to the grocery cooperatives and restaurants with his trips to the farmers markets. He was relying on Fiddler's 14-year-old box truck (unrefrigerated) with 270,000 miles to transport his produce, but it now needs a new engine, so Eldon is currently driving to his markets in his pick-up truck, which averages 12 mpg. On Saturdays, he first delivers to the Davis Food Cooperative and then drives less than a mile to the Davis Farmers Market (35 total food miles). After the market, he makes deliveries to a

Davis restaurant, Tucos Wine Bar and Café, and the grocery cooperative in Sacramento, driving another 20 miles into Sacramento, and 50 miles back to the farm; the roundtrip totals 105 miles. Food miles to the Marin Farmers Markets and restaurant customers total 85 miles. Eldon's loads vary from 800 pounds to 2,000 pounds, for an average of 1,400 pounds. Thus fuel usage averages 63 gallons and 1.04 gal/cwt of produce for the two market routes.

Community and Economic Linkages

Eldon noted that he benefited significantly from participating in the Community Alliance with Family Farmers' Lighthouse Farm Network, which was designed to connect farmers and other agricultural professionals who meet once a month to share technical information about biologically based farming practices. Despite his busy schedule, Eldon has been generous in sharing his time with civic organizations. He served on the Marin Farmers Market Board of Directors for ten years and one term on the board of the Community Alliance with Family Farmers. He is now in his fourth year as a member of the Davis Farmers Market Board of Directors, where he serves as the treasurer. Eldon was also on the Certified Farmers Market Advisory Committee for California's Department of Food and Agriculture for two terms. He supports food programs for low income residents by donating unsold produce from his farmers market loads to two community nonprofits, Short Term Emergency Aid Committee (STEAC) in Davis and Meals of Marin in San Rafael.

Fiddler's is a full-time business for Eldon. His wife Julie is employed full-time off the farm. Fiddler's payroll expenses total approximately $90,000 for the four employees who live close by. Eldon purchases his fuel for the farm locally as well as his farm implements. His pump and cooler maintenance service providers are also local. Since Fiddler's sells only fresh produce, it does not rely on any processors. Overall, Fiddler's economic impact on the local economy is limited because of its relatively small size. If Fiddler's were to cease operations, another farming operation would be likely to take over farming the land.

Prospects for Expansion

After becoming a partner, Eldon expanded Fiddler's rapidly and then suffered significant losses when the freeze hit his plantings in the spring of

1999. He was forced to discontinue his CSA program and has not restarted it. He is concerned that because of losing his source for transplants last spring, his farm will not be profitable in 2009. He has considered starting up another CSA, since farmers typically earn full retail prices for their CSA programs while incurring lower marketing costs than at farmers markets. However, Eldon is hesitant to do so for two reasons: (1) it would involve considerable administrative effort to develop and operate the CSA; and (2) he would face considerable competition, since at least nine farms within 25 miles of Fiddler's currently operate CSAs. Alternatively, Eldon could consider a hybrid form of CSA, perhaps for a major employer in Marin County, that could be piggybacked to his trip to the Thursday Farmers Market.

Eldon is an effective marketer at farmers markets. His produce is attractively displayed at his stall, and he is approachable and engaging. Eldon's best prospects for renewed profitability appear to be with expanding into new crops on his existing acreage. Fiddler's earns a considerable premium for its spring mix at the Davis Farmers Market during the summer and early fall when no other farmers have it. His challenge is to find similar niche crop(s) that can command premium prices to bolster his revenues.

Eldon's major concerns with policy issues are the time it takes to understand new regulations and the recordkeeping effort required to comply with them. He does not have any administrative support and must find the time to complete all required reports himself. When asked specifically about potential food safety regulations, Eldon did not express much concern. He commented that he does not feel vulnerable to a food safety outbreak with his leafy greens because they are a "loose leaf" product, rather than packaged.

Key Lessons

This case study of Fiddler's demonstrates three general lessons about direct marketing. First, it exemplifies the synergies in distribution possible from marketing through different channels. When Eldon travels to a geographic market area, he is selling through two or three channels, which means that his marginal marketing costs in the secondary channels are minimal. During the summer Eldon may be able to enhance Fiddler's profitability by producing and selling more spring mix at the Davis Farmers Market when he is the only spring mix marketer.

Fiddler's situation also illustrates that there can be substantial competition in direct marketing channels. Specifically, Eldon has to sell his spring mix at the Marin Farmers Market for $6.00/lb instead of the $8.00/lb he earns at the Davis Farmers Market, where there are no other farmers selling high quality product over a long period. Similarly, Eldon senses that his prospects in relaunching his CSA program could be difficult since there are several local organic farmers who already have strong CSA programs. While the growth in direct marketing has enabled many smaller farmers to enter the marketplace, farmers markets, CSAs, and other forms of direct marketing often now involve significant competition among farmers selling to a relatively small number of customers.

Additionally, Fiddler's situation exemplifies the vulnerability faced by smaller producers who rely heavily on directly marketing. Fiddler's has reduced its marketing risk significantly by growing a very diverse set of crops. However, such diversification did not protect Fiddler's from the catastrophic impact of the 1999 spring freeze; the farm's revenues plunged from $500,000 to under $150,000 as it closed its CSA program and reduced its acreage. Though not as severe, the lack of tomato and pepper transplants in the early spring also damaged Fiddler's significantly in 2009, as did the breakdown of the box truck that Eldon has relied on for so long to haul his produce. For a larger farming operation, the loss of a single supplier or a single vehicle is not likely to have such a severe impact. However, as a small direct marketer, it is essential for Fiddler's to have a reliable delivery vehicle to travel to its markets, and to be able to market as wide of a range of produce as early (and as late) in the season as possible, in order to earn premium prices when product availability is limited.

Intermediated Case: Davis Food Coop

Davis Food Cooperative (referred to as the Co-op) is owned by approximately 10,000 households in Davis. The university-oriented community is located 15 miles west of downtown Sacramento. The full-service grocery store (davisfood.coop) carries approximately 18,000 products and features a sushi vendor, hot food to go, hot soup, salad bar, custom sandwiches, and store-baked pastries.

The Co-op is governed by a 10-member board that includes one store employee. The board recently adopted a "statement of ends" (similar to a mission statement) stating, "We are the best source of healthful, sustainable, higher quality, and locally grown and produced foods." In writing about this statement in a recent newsletter, the Co-op's general manager further noted: "Buying from local growers makes sense for any number of reasons, including flavor, freshness, reduced transportation, and preservation of local farms" (Stromberg 2009: 2). Rather than trying to compete on a price basis with other grocery stores, the Co-op management's strategy is to know its customers and to understand their needs.

The Co-op has operated at the same location for over 25 years and it is open seven days a week. The 25,000-square-foot store is large by grocery co-op standards but considerably smaller than the median U.S. grocery store of 46,755 square feet (Food Marketing Institute 2009). It is completing an extensive remodeling project that began in 2006. In 2008 its revenues totaled $18.1 million, making it the third largest cooperative grocer in California. The Co-op does not have a warehouse, but rather, numerous suppliers deliver their products directly to the store.

The Co-op plays a critical role as the retail hub for consumers, ordering from numerous sources and selling the product. Its produce department carries more than 900 items during the year, and more than half of the items are organic. Produce sales for the 2008–2009 fiscal year totaled $3 million, and approximately 80 percent of produce sales are organic. It is staffed by 13 employees; nine are full-time and four are part-time. Elizabeth Davidson has been the Co-op's produce manager for the past 15 years. When placing orders, she does not use any forecasting models, instead relying on her past experience. The Co-op does not have any contracts with its produce suppliers.

Although California is the nation's leading supplier of spring mix, a distinct feature of this intermediated spring mix supply chain in the Sacramento area is that the local producers are not part of the Salinas Valley, which is approximately 175 miles away. Rather, these farmers grow numerous leafy greens along with an extensive mix of other vegetable crops and fruit crops. Several also engage in livestock and/or poultry production. All of the spring mix, head lettuce, and leaf lettuce sold by the Co-op is organic. Spring mix is sold in bulk and clamshell packages.

Supply Chain Structure, Size, and Performance

The Co-op's supply chain for spring mix is shown in figure 15. The store sources spring mix from suppliers with three distinct organizational profiles: four local farms that supply bulk spring mix, a distributor that supplies bulk spring mix, and a large distributor that supplies only packaged spring mix. All these sources supply other produce items to the Co-op as well. The Co-op strives to source bulk spring mix exclusively from local farms when it is available. When locally grown spring mix is not available, the Co-op purchases bulk spring mix from a San Francisco-based produce distributor, Veritable Vegetable. Earthbound's spring mix and other salad mix products in clamshell packaging are sourced year-round from a local produce distributor, Nor-Cal Produce (also the sole distributor for the mainstream case, Nugget Markets). These three supply chains, along with their ordering, delivery, and payment procedures, are described in the following sections.

Production—Local Spring Mix

The Co-op purchases locally grown spring mix from four farms all located in Yolo County, which has the highest agricultural production of the four counties in the Sacramento area. All four local farms have coolers onsite for holding their harvested produce. The locally grown product represents, at most, 1 percent of the Co-op's spring mix sales. The Co-op is not the primary market for spring mix grown by local growers. Three of the local farms are approximately the same size. They market spring mix during the spring, late fall, and early winter, if weather conditions are appropriate. During the fall, hot weather can ruin the crop. Production during the winter and spring is adversely impacted by lack of sunshine, heavy rains, and mildew. These three local farms market their spring mix primarily through their CSAs. Although located within 20 miles of the other farms, the fourth farm—Fiddler's Green—has unique climatic conditions that enable it to extend its spring mix production through much of the hot summer. More information about Fiddler's Green Farm is presented in the preceding direct marketing case.

Thus local spring mix is marketed to the Co-op for a relatively short period, even when weather conditions are favorable. During periods of short supply, local growers often keep spring mix on their wholesale

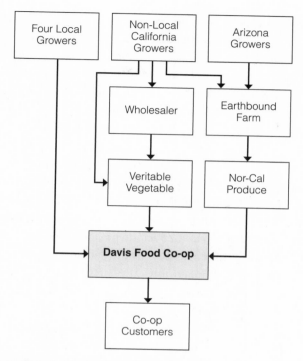

Fig. 15. Intermediated supply chain, Davis Food Co-op.

product availability lists but price it high in order to keep the cooperative's purchases low.

The Co-op's produce manager strives to rotate her orders among the local growers. However, due to growing conditions and pricing, Terra Firma is the Co-op's primary local spring mix supplier. It was founded in 1985 by one of the three current partners, Paul Holmes. The second partner, Paul Underhill, joined about ten years later. A third partner, Hector Melendez, was added about three years ago after having worked for the farm for many years as a farmworker. Terra Firma is highly diversified. It operates year-round on 240 acres and grows approximately 60 crops annually, including fruits, vegetables, nuts, and grains. It has been selling spring mix to the Co-op for about 20 years.

Terra Firma is structured as an S-corporation, a legal ownership form for small businesses with a small number of shareholders that is exempt from corporate income tax. For most of the year it has 35 full-time employees. It brings in an additional 130 employees for three weeks each year to hand-harvest a small grain that it markets as birdseed. Terra Firma earns approximately 40 percent of its revenues from its 1,400 CSA memberships. Although farmers markets used to be a significant sales outlet for Terra Firma, it discontinued its farmers market program with the exception of one fledgling local market that generates less than 1 percent of its revenues. Terra Firma earns about 15 percent of its revenues from wholesale sales to the Co-op, another grocery cooperative in the Sacramento area, and the Whole Foods store in Sacramento. Restaurants account for approximately 5 percent of Terra Firma's revenues. The remaining 40 percent of Terra Firma's revenues come from sales to distributors, including two in the Sacramento area. None of Terra Firma's restaurant or distributor customers in the Sacramento area or the Bay Area for its numerous products buys any spring mix from Terra Firma; these customers can source product grown in the Salinas Valley that has more consistent quality and a longer availability period.

Terra Firma is known for its heirloom tomatoes, which it also sells to the Co-op. It generates approximately 2 percent of its revenues from spring mix and does not consider spring mix to be a highly profitable crop. Rather, it grows spring mix because this is a popular item with its CSA members during the fall and winter months; half-pound bags are included in CSA boxes. Terra Firma grows all its spring mix greens from seed. The mix includes a variety of baby lettuces—red perella, red oakleaf, red leaf, and green romaine along with frisee, spinach, radicchio, and arugula. It starts the frisee and radicchio in a greenhouse.

Terra Firma harvests the greens in its spring mix by hand early in the morning. The harvested greens are transported to a packing shed, where they are washed together in tubs that have been sterilized with bleach. After washing, the greens are dried in mechanical salad spinners that hold approximately eight pounds of spring mix. The spun greens are packaged in perforated salad bags, which are placed into four-pound boxes for delivery to the two grocery cooperatives and eight-pound boxes to be bagged for CSA subscribers. The boxed product is placed into a refrigerated truck and hauled three miles to one of Terra Firma's coolers. The spring mix is stored at 38 degrees in the cooler.

Holmes noted that Terra Firma stopped growing spring mix during January 2009 because of heavy rains. They began replanting the lettuces in early September 2009 but had to wait until early November to harvest spring mix for the CSA boxes due to 100-plus temperatures in September followed by heavy rains in early October, ruining Terra Firma's first two spring mix plantings.

Distribution

The delivery process is the same for all suppliers. The deliveries are on pallets, making unloading and storage easier for the Co-op. All the spring mix products are delivered to the store's back door, unloaded, and moved into the Co-op's 32-foot by 12-foot walk-in cooler, where it is stored at 34 to 40 degrees, usually for no more than two days. Spring mix is a high rotation item. The Co-op's payment process is identical for all its suppliers. All the suppliers provide an invoice when making a delivery. Every load is inspected by the Co-op upon delivery. Outright rejections are rare for spring mix, because the Co-op's produce manager emphasizes the importance of advising her of quality and quantity problems ahead of the delivery. The Co-op pays all its suppliers, including local farmers, by check within 14 days of the delivery. Occasionally other local growers approach the Co-op's produce manager, wanting to sell produce to the store on a cash basis; she explains the Co-op's purchasing and payment policies to them.

Distribution—Local Spring Mix

The local farmers email availability lists to the Co-op's produce manager daily. She compares the prices for spring mix when more than one local grower has the product available, since one farm may offer a sale price because they have excess supply. Terra Firma's price for spring mix is typically slightly lower than those of the other three growers. The produce manager phones in her order for spring mix, along with other produce items, to one of the growers twice a week. Terra Firma and the other local growers advise the Co-op's produce manager about any quality and quantity problems with the spring mix during the call, which she remarked is much better than receiving a "shorted" order and/or poor quality product. In the fall of 2009 Terra Firma suddenly stopped selling spring mix to the Co-op because its butternut squash crop had failed, and it substituted larger quantities of spring mix in its CSA boxes. If the quality of Terra Firma's spring mix is poor, the Co-op asks the other farms about the quality of their spring mix and places an order with one of them if the quality is good. If none of the local growers has good quality spring mix, she orders it as a nonlocal product from a San Francisco-based organic produce distributor, Veritable Vegetable.

The local farmers are all within 45 miles of the Co-op. Terra Firma delivers to the Co-op as part of its local delivery route in a diesel refrigerated truck. On Tuesdays the truck delivers to the other local grocery cooperative, another local grocery store, and two local distributors. On Fridays the truck also delivers to the other local grocery cooperative and the local CSA drop sites. Terra Firma does not itemize a delivery charge in its invoices. Terra Firma and the other local farmers usually make their deliveries to the Co-op during the early afternoon on Tuesdays and Fridays.

Distribution—Bulk Nonlocal Spring Mix

Veritable Vegetable supplies the Co-op with nonlocal bulk spring mix as well as other organic produce items. Founded in 1974, Veritable Vegetable is one of the pioneers in the organic produce industry and the nation's oldest distributor of certified organic produce. Its distribution area includes California, New Mexico, Arizona, and Colorado; however, 75 percent of its sales are attributable to customers in California. Veritable Vegetable's 300 customers include retailers, restaurants, and other regional distributors. It operates a 25,000-square-foot warehouse facility containing more than 9,700 items, of which 97 percent are certified organic. It buys from 340 produce vendors and employs more than 80 full-time staff.

Veritable Vegetable's mission is to understand and provide for the unique needs of each of its customers. The Co-op has been buying produce from the firm for 30 years. Its primary supply period for bulk spring mix is February through September; however, the Co-op also expects Veritable Vegetable to be a "back-up" supplier during the local spring mix season. When it is available, Veritable Vegetable purchases bulk spring mix grown in California's Salinas Valley. During the "off-season" it sources spring mix grown in California's Imperial Valley and Yuma, Arizona; however, this spring mix is ordered from Salinas Valley processors, so shipments to Veritable Vegetable come from the Salinas Valley. Some of Veritable Vegetable's bulk spring mix is supplied by Earthbound, the firm that processes the packaged spring mix products sold by the Co-op that are purchased through a different distributor. Detailed information about Earthbound's production, harvesting, processing, and transportation practices are provided in the mainstream case featuring Nugget Markets.

Veritable Vegetable emails availability lists daily to the Co-op's produce manager, who in turn submits orders to the firm four days a week. Veritable Vegetable makes the deliveries the next day and has the capability to deliver to the Co-op six times a week. Deliveries are made in a 48-foot tractor-trailer. The truck is usually routed to go farther east to Grass Valley for 250 roundtrip miles or farther north to Chico for 450 roundtrip miles.

Distribution—Packaged Spring Mix

The Co-op also purchases packaged nonlocal spring mix year-round (as well as other produce) from produce distributor Nor-Cal, located in the Sacramento area. Nor-Cal's warehouse is 16 miles from the Co-op, making it the Co-op's closest produce supplier. Its sourcing and transportation practices are described in detail in the Nugget case study. The Nor-Cal sales representative to the Co-op used to work in the Co-op's produce department. The Co-op's produce manager receives a product availability list via email each day then phones in her order to Nor-Cal and relies on her salesperson to advise her of any unusual supply conditions. Nor-Cal delivers to the Co-op six days a week.

Virtually all the Co-op's packaged spring mix is the Earthbound brand, which is headquartered in the Salinas Valley. The Co-op began purchasing Earthbound's packaged salads in clamshells in January 2007 when its new refrigerated produce case was installed. In 2009 the order size averaged 50 cases for a variety of Earthbound salad products, including the spring mix. Previously the Co-op purchased a different salad brand packaged in bags. Nor-Cal also supplied the Co-op with another brand, Organic Girl (also headquartered in Salinas Valley), for two weeks during 2009. The Co-op's produce manager discontinued the Organic Girl product after deciding the Earthbound product was of higher quality.

Marketing and Retail Operations

The Co-op displays the bulk spring mix with other organic leafy greens. The nonlocal product is labeled USA. Sometimes, the bulk spring mix has been processed by Earthbound; however, no brand information is included in the bulk display. When local spring mix is available, the sign indicates "Local/CA" because there is always a possibility that the local product will sell out and only nonlocal bulk spring mix is available until the next delivery by a local grower. Currently the Co-op is carrying Earthbound's five-ounce and one-pound clamshell packages of spring mix as well as 12 other Earthbound packaged salad products.

Nor-Cal has locked in pricing on all of Earthbound's spring mix products sold to the Co-op. Nor-Cal receives ad discounts each week from Earthbound that rotate between the different sizes. The discount is $1.00 on the three-pound bulk product, $1.20 on the case of eight five-ounce packages, and $3.00 on the case of six one-pound packages. The Co-op runs ads frequently on packaged salads. The store has the discretion to keep all or part of the promotional allowance to meet its margin targets.

The Co-op's average retail prices per pound and sales volumes for the different spring mix products are displayed in table 11. Its sales of spring mix as a proportion of total produce sales range from 5 percent to 15 percent per month, peaking during the spring and summer. Based on actual sales through November 22, 2009, sales of spring mix for the entire year were projected to total $48,427 and 8,753 pounds. By weight, almost two-thirds of the Co-op's spring mix is sold in bulk, and 26 percent is attributable to the one-pound product. This proportion, however, drops to 28 percent on a retail dollar volume basis. The per pound price of the five-ounce package is almost double that of the one-pound package, and the price of the bulk product is 21 percent lower than that for the one-pound package.

The produce manager perceives strong demand for local produce in general; however, demand for local spring mix is not particularly strong. Consumers have complained to her that the product does not seem to last as long as the packaged spring mix they purchase. The produce manager remarked that "they don't understand that it is not the same product," and she noted that Earthbound's bulk spring mix has a very consistent mix of greens and a longer shelf life.

Table 11. Co-op's spring mix prices and sales, 2009

PRODUCT	AVERAGE PRICE/LB	POUNDS SOLD[a]	SALES	% OF TOTAL LB SOLD	% OF TOTAL $ SALES
Earthbound, 5-ounce	$11.29	683	$7,705	7.8	15.9
Earthbound, 1-pound	$5.90	2,265	$13,365	25.9	27.6
Bulk, local	$5.99	100	$599	1.1	1.2
Bulk, not local	$4.69	5,705	$26,757	65.2	55.3
Total		8,753	$48,427	100.0	100.0

a. Projected, based on sales through November 20, 2009.

Source: Case study interviews.

As a consumer-owned grocery cooperative, the Co-op is committed to purchasing from local growers as part of its mission of supporting the local community. Each year the Co-op's management visits one of the local farms and has a lunch meeting with the growers. The Co-op's produce manager has established good working relationships with local farmers during her 15-year tenure. She believes that the communications between her and the local farmers regarding the quality and availability of their produce are essential for having a successful relationship. The other half of this values-based relationship is that she understands why there can be intermittent disruptions in the supply of locally grown spring mix. Such disruptions would not be acceptable to a large grocery chain.

Food Miles and Transportation Fuel Use

As indicated in table 12, food miles for Earthbound's packaged spring mix totaled a weighted average of 414 miles for spring mix grown in the Salinas Valley and in the desert (detailed calculations are reported in the Earthbound case). Loads averaging 13,000 pounds are hauled from the farm to Earthbound's plant with an average distance traveled of 36 miles. Hauling product from the Earthbound plants to Nor-Cal's

Table 12. Spring mix food miles and fuel usage for Davis Food Co-op's supply chains

SUPPLY CHAIN	FOOD MILES TRAVELED	ROUND-TRIP MILES
Packaged Spring Mix—Blended for Earthbound's Salinas and Desert Production		
Farm to Earthbound	36	72
Earthbound to Nor-Cal	362	723
Nor-Cal to Co-op	16	150
Total	414	945
Bulk Nonlocal Spring Mix—Blended for Salinas and Desert Production		
Farm to processor	36	72
Processor to Veritable Vegetable	304	603
Veritable Vegetable to Co-op	75	350
Total	415	1,025
Bulk Local Spring Mix		
Farm to Co-op	22	95

Source: Case study interviews.

warehouse is done in a 50-foot refrigerated truck and involves the greatest fuel use. The Co-op is only 16 miles from Nor-Cal's warehouse, but the entire delivery route is 150 miles for the truck with a 48-foot trailer. Given the roundtrip nature of the hauling and the 6 mpg fuel efficiency, the weighted average total fuel used for all legs of this supply chain totals 0.52 gal/cwt of spring mix.

Transportation for the nonlocal bulk spring mix supply chain involves the same transportation as for Earthbound product for the first two legs. The third leg, from Veritable Vegetable's warehouse in San Francisco to the Co-op in Davis as part of a longer route to Chico, increases the food miles from 16 to 75 and roundtrip miles from 945 to 1145. Consequently,

MILES/ GALLON	GALLONS/ LOAD	LOAD (100 LB)	GALLONS FUEL USE/ 100 LB
6	12.0	130	0.09
5.5	131.5	400	0.33
6	25.0	250	0.10
			0.52
6	12.0	130	0.09
5.5	110.5	400	0.28
6	58.3	250	0.23
			0.60
10	9.5	60	0.16

total fuel usage in the nonlocal bulk spring mix supply chain rises to 0.60 gal/cwt, assuming six mpg fuel efficiency for most of the miles.

The local grower clearly provides the shortest supply chain and travel. The food miles traveled drop to 22 miles and fuel use for the 16-foot refrigerated box truck (averaging fuel efficiency of 10 mpg) is 0.16 gal/ cwt of spring mix, resulting in 73 percent less than the highest fuel usage of 0.60 gal/cwt for the nonlocal bulk spring mix supply chain. Low food miles do not always translate into low fuel use; the local grower achieves significant fuel efficiency by using a relatively large truck that includes deliveries of various types of produce at CSA drop sites and other grocers as part of the delivery route.

Community and Economic Linkages

The Co-op is engaged in a variety of community activities, none of which is specifically related to spring mix. It offers a variety of cooking classes and holds wine and beer tastings that bring community members together. Holiday celebrations include breakfast with the Bunny, a haunted house outside the store, and breakfast with Santa. Its monthly newsletter includes grower profiles as well as healthy recipes.

One of the Co-op's principles is "Concern for the Community." The Co-op provides substantial support to the local community, such as educating local school children about whole grains and eating seasonally, handing out recipes and samples of seasonal foods at a farmers market, and organizing community traditions such as the Children's Candlelight Parade and the Holiday Meal (which raised over $14,000 in 2008). They also sponsor various community programs and provide grants to support various charities, including the local Society for the Prevention of Cruelty to Animals, Land Trust, Red Cross, Youth Soccer League, and the campus radio station.

Terra Firma's founding partner, Paul Holmes, remarked that the Co-op's willingness to buy from local, smaller producers has given them the opportunity to establish a track record which they can then use to acquire more retail customers. The Co-op encourages local producers to participate in store sampling events for their products, giving the producers an opportunity to improve their marketing skills with consumers.

As a consumer grocery cooperative, the Co-op differs from national grocery chains and regional grocery markets because it is owned by its customer-members. Member equity of $2 million accounted for 29 percent of the value of the cooperative's assets, and retained earnings totaled $1.9 million in 2008. Although it is not currently doing so, the Co-op has paid out patronage refunds to members in the past. Also, member households can receive a 5 percent discount on their purchases for each month they work (two hours for a one- to two-person household).

As part of the Co-op's commitment to being a leader in the local foodshed, the store makes a special effort to market foods and other products produced within 100 miles from the store. Products that are locally produced have a special local shelf tag. The Co-op's payroll totaled $4.3 million in 2008 for its 130 employees. Given the employees' concern for sustainability, it is likely that a majority of them live in the local area and spend a significant portion of their earnings locally. In summary, the Co-op's community and economic linkages are significant; however, they clearly cannot be attributed solely to its spring mix supply chain.

Prospects for Expansion

While there is significant interest in local foods among the Co-op's customers and the Co-op has a strong commitment to sourcing produce and other food products from local suppliers, prospects for expansion of the Co-op's sales of locally grown spring mix do not appear to be strong. Barriers to expansion include local growers' high production costs relative to those who machine harvest large plantings of spring mix, causing local growers to favor marketing directly to consumers and selling to the Co-op only when they have excess supply of spring mix. Second, weather conditions do not favor a consistent supply of high quality product that Salinas Valley growers can produce from spring through late fall. The area's summer temperatures are generally too hot, and fall plantings can be ruined by heat waves as well as heavy rains.

Compliance and Food Safety

The Co-op has not encountered any major regulatory problems related to its local food supply chains. As a major purchaser and seller of certified organic produce, the Co-op has not had any difficulties with the USDA's National Organic Program. The produce manager asks all her suppliers to provide their organic certificates annually. California's Department of Food and Agriculture has packaging standards for all produce that is purchased for resale. The Co-op was purchasing from a local grower who was packing produce in nonstandard boxes; it stopped buying from the grower until the grower got new boxes.

The Co-op's produce manager did not express much concern about food safety standards; the Co-op has never been associated with a food safety outbreak related to produce or any other foods. However, its supplier of packaged spring mix salads—Earthbound—was identified as the processor of the contaminated baby spinach that resulted in the highly publicized 2006 *E. coli* 0157:H7 outbreak. Terra Firma's Paul Holmes expressed great concern about food safety regulations for leafy greens.

Currently the Co-op does not require its four local spring mix growers to be compliant with the provisions of the California Leafy Green Marketing Agreement (LGMA), and none of these growers supply any of the California LGMA's signatories.

Holmes indicated that if Terra Firma was required to comply with the National LGMA, it would no longer grow any leafy greens. Recent research indicates that costs of compliance with California's LGMA and other food safety standards approach $100/acre, with smaller producers having higher per acre costs than larger producers (Hardesty and Kusunose 2009). Additionally, FDA is initiating efforts to develop nationwide food safety standards for all produce.

Key Lessons

Spring mix is one of the many locally grown produce items sold by the Co-op. Spring mix sales, however, are very limited because local growers prefer to market most of their product direct to consumers. They earn higher returns by selling spring mix through farmers markets and their CSA programs, although the Co-op is paying them a significantly higher price than Salinas Valley growers are paid by Earthbound and other salad processors. The Co-op is willing to earn lower margins on the locally grown product in order to keep it priced competitively.

The Co-op's produce manager incurs significant transaction costs to source locally grown spring mix. The available supply to the Co-op is erratic during the fall season due to both weather conditions and the growers' treatment of the Co-op and other local grocers as a residual market. The Co-op often runs out of product and has to have a back-up supply of "mainstream" product from a distributor. Additionally, the Co-op produce manager is sometimes asked to raise the retail price for the local spring mix such that it is on par with the price the growers charge at farmers markets. Nevertheless, she has a strong relationship with the four local growers, due largely to the other vegetables and fruits that these growers sell to the Co-op.

The Co-op is a particularly interesting illustration of the market for locally grown produce because it uses three different supply chains for spring mix in order to have a reliable supply of this delicate and high-volume produce item. Locally grown spring mix represents only 1 percent of the Co-op's sales (weight-wise). As described, however, its availability is limited and consumer demand for it does not seem to be particularly strong. Packaged spring mix makes up one-third of its spring mix sales (weight-wise), and part of its supply chain is the same one examined in the mainstream case. Bulk spring mix that is not locally grown is the Co-op's primary spring mix product. It is supplied by a long-time distributor of organic produce that sources spring mix from several smaller processors operating in both the Salinas Valley and the desert.

Cross-Case Comparisons

The three Sacramento area case studies conducted—Nugget Markets, Fiddler's Green Farm, and Davis Food Co-op—demonstrate the variety of market sources for spring mix available to consumers. When compared against each other, they illustrate a complex array of differences in supply chain structure, size, and performance across supply chain types.

Supply Chain Structure

In reviewing the structure of local food supply chains, the findings highlight, for the most part, structural differences between the two local food supply chains and the mainstream chain.

Consumers receive the most detailed information about where, how, and by whom their spring mix is produced through the direct market local food supply chain. Although the Co-op has a well-developed local food program, spring mix is one of the rare produce items the Co-op does not identify as "local spring mix," and it does not provide an individual farm name, because it frequently runs out of the local product. Instead the Co-op labels local spring mix as "Local/California." Spring mix is not part of Nugget's relatively limited local produce marketing program.

Durable relationships between supply chain partners are evident across all chains. There is significant information exchange and trust between mainstream supply chain members (Nugget and Nor-Cal) and between the intermediated supply chain members (the Co-op and its local grower suppliers). Similarly, in the direct market supply chain, Fiddler's has loyal customers at the farmers market who trust Eldon to provide them with safe and fresh product.

As expected, the prices charged by Fiddler's, Terra Firma, and other local growers who supply the Co-op in the intermediated supply chain are decoupled from commodity markets. Since the mainstream supply chain member, Earthbound, is the brand leader for organic spring mix (as well as being a major supplier of private label spring mix), we can conclude that Earthbound has major influence on the commodity price of spring mix. Information on prices paid to Earthbound's growers was not available; however, we surmise that to ensure steady supplies, they have season-long contracts with Earthbound, paying a stable price.

Collective organizations, particularly farmers markets, have contributed significantly to the success of local supply chains for spring mix. Currently Fiddler's is generating 70 percent of its revenues from sales at farmers markets. When the local producers first began marketing their spring mix and other produce, the farmers markets served as a marketplace where they could earn a premium for their organic produce, provided them access to wholesale customers as well as consumers, and created the initial customer base for their CSA programs. Although consumer grocery cooperatives serve as important intermediaries in the marketing of local produce, their role in marketing spring mix has been limited, due to the higher prices earned by producers in direct market channels.

Fiddler's and the other local spring mix growers who market through direct and intermediated supply chains are not linked to the national industry infrastructure that is based only 175 miles away. The production and handling methods of the local growers are vastly different from those of the larger and more specialized producers in the Salinas Valley.

Local spring mix producers have benefited significantly from the strong local food infrastructure provided by the farmers markets, consumer grocery cooperatives, and CSAs in the Sacramento area. As noted, direct marketer Eldon developed new restaurant customers through contacts at the Davis Farmers Market. Both he and Terra Firma gained significant transportation efficiencies by being able to combine deliveries to multiple local customers within a single trip.

Supply Chain Size and Growth

Size differences among these supply chains are noticeable. Fiddler's markets about 2,000 pounds of spring mix annually, while the Co-op's volume totals approximately 8,800 pounds with only 100 pounds of local product. By contrast, Nugget's spring mix sales in the mainstream supply chain average approximately 6,500 pounds per store annually, and none of it is local.

Access to processing and distribution is not a significant barrier to expansion for Fiddler's or the other local growers supplying the Co-op. However, Earthbound does benefit from significant scale economies associated with its mechanized harvesting and processing.

Fixed costs for compliance with regulatory and operating standards limit the potential size of the chains. Following recent outbreaks of food-borne illness, food safety operating standards have been broadly adopted by leafy greens handlers supplying mainstream markets. Thus far, their impact on the smaller local producers has been negligible because these producers have not sought distribution in these markets.

The FDA, however, issued a draft food safety guidance document for leafy greens in 2009. In December 2010 Congress passed food safety legislation that exempts small farmers who direct market the majority of their production. Meanwhile, the FDA still needs to issue its regulations and obtain funding for its enforcement costs. If some form of good agricultural practices standards is eventually adopted that imposes regulations on all growers, the high compliance costs could make spring mix production unprofitable for smaller growers in local supply chains. Additionally, the USDA issued a proposal in April 2011 for a National Leafy Greens Marketing Agreement (USDA-AMS 2011b). Modeled after the California Leafy Greens Marketing Agreement, it would establish a voluntary food safety program with uniform standards nationwide for the participating handlers; these handlers, in turn, would require compliance from the growers who supplied them with leafy greens.

Lack of year-round availability limits market opportunities for local spring mix. Although spring mix was available at the Davis Farmers Market during 48 of the 49 weeks when we collected data for this project, supplies were limited during half of the year. This limited availability creates a thin market with high prices for local spring mix and restricts supplies in the intermediated supply chain where wholesale prices of nonlocal spring mix are significantly lower.

Expansion opportunities are mixed across the supply chains. For example, in the direct market supply chain, expansion is likely to come through entry of new growers at farmers markets. However, growth in the intermediated supply chain is unlikely because local growers earn higher prices by direct marketing their spring mix.

Supply Chain Performance

Allocation of retail revenue for spring mix varies widely across the three supply chains. Many growers are attracted to farmers markets because they can sell their produce at retail prices or even earn a premium over the regular retail price. When adjusted for marketing costs, the producer's share of revenues decreases with distance to market and the number of intermediaries involved in the supply chain (table 13). Eldon's costs for marketing at the farmers market represent an estimated 26 percent of the revenues he generates. Eldon incurs marketing expenses selling at farmers markets, such as transportation costs, farmers market stall fees, and labor costs for driving to and from the market, setting up the stall, staffing it, and taking it down. Even if Eldon is performing the work himself rather than hiring someone, he is incurring opportunity costs for his time. Nevertheless, selling at the farmers markets also gives him the opportunity to talk to current and potential customers, in order to gain better understanding of their needs, get their feedback about his produce, and potentially develop new restaurant customers.

Median prices for bulk spring mix over 2009 varied considerably. The median price per pound at the farmers market was $6.00 for conventional, $8.00 for Fiddler's organic product, which is available primarily during the local off-season, and $7.00 for other local growers. Fiddler's also earned a significant premium over nonlocal spring mix. Its price was 71 percent higher than the Co-op's $4.69 median price, and 23 percent higher than Nugget's consistent $6.49 price. These data suggest that consumers are willing to pay more for three attributes—organic, off-season, and direct availability from a local producer.

Revenue retention within the local economy appears to be relatively high in all three supply chains. Fiddler's owner Eldon, and the other local spring mix growers and Fiddler employees, live locally and presumably spend most of their household earnings in the Sacramento area. The Co-op has approximately 130 full-time and part-time employees, with payroll expenses

Table 13. Allocation of retail revenue in Sacramento, California— spring mix chains, by supply chain and segment

Supply chain Segment	MAINSTREAM Nugget Market Revenue ($/lb)	% of total	DIRECT Fiddler's Green Revenue ($/lb)	% of total	INTERMEDIATE[d] Davis Food Co-op Revenue ($/lb)	% of total
Producer[a]	0.79	12.2	5.92	74.0	3.00	50.1
Producer estimated marketing costs[b]	0.02	0.3	2.08	26.0	0.75	12.5
Processor	1.16	17.9	na	na	na	na
Distributor[c]	0.77	11.9	na	na	na	na
Retailer	3.75	57.8	na	na	2.24	37.4
Total retail valued	6.49	100	8.00	100	5.99	100

For the direct and intermediated supply chains, the farm also operates as the processor.

[a.] Mainstream: Calculated as a weighted average of farm gate prices paid in Monterey and Imperial Counties, 60 percent and 40 percent, respectively, and adjusted for 45 percent of the production in each county earning a 10 percent price premium for organic product. Direct and Intermediated: Includes compensation for processing activities, such as washing, mixing, and bagging.

[b.] Mainstream: Includes estimated costs of transportation to the processor. Total farm per unit revenue is 0.79 + 0.02 = 0.81 ($/lb). Direct: Includes estimated transportation costs, farmers market stall fees, and opportunity costs of time for marketing activities. Total farm per unit revenue is 5.92 + 2.08 = 8.00 ($/lb). Intermediated: Includes estimated transportation and packaging costs. Total farm per unit revenue is 3.00 + 0.75 = 3.75 ($/lb).

[c.] Includes compensation for inbound freight charges averaging $0.50/pound for bulk spring mix.

[d.] Mainstream and Direct: Median retail price of bulk spring mix from January to December 2009. Intermediated: Median retail price of bulk spring mix from January through March 2009.

Source: King et al. (2010: 32).

Table 14. Food miles and transportation fuel use in Sacramento, California—spring mix supply chains

SUPPLY CHAIN SEGMENT	FOOD MILES	TRUCK MILES
Mainstream: Nugget Market (CA)		
Producer to Processor-Shipper[a]	30	60
Processor-Shipper to Distribution[b]	192	372
Distribution to Retail[c]	16	150
All Segments	238	
Mainstream: Nugget Market (AZ)		
Producer to Processor-Shipper[a]	45	90
Processor-Shipper to Distribution[b]	618	1,250
Distribution to Retail[c]	16	150
All Segments	679	
Mainstream: Nugget Market (CA & AZ combined)		
All Segments[d]	414	
Direct: Fiddler's Green		
Producer to Retail[e]	35	105
All Segments	35	
Intermediated: Davis Food Co-op[f]		
Producer to Co-op[g]	22	95
All Segments	22	

[a.] These short-haul loads use a trailer that achieves fuel economy of 6.0 mpg.

[b.] These loads are transported in a tractor-trailer that achieves fuel economy of 5.5 mpg.

[c.] These loads are transported in a tractor-trailer that achieves fuel economy of 6.0 mpg.

[d.] Food miles and fuel use per cwt are calculated as the average of the CA and AZ chains, weighted by the total product weight in each chain (60% for CA, 40% for AZ).

[e.] All transport in this chain is in a box truck that achieves fuel economy of 12 mpg.

RETAIL WEIGHT (CWT)	FUEL USE (GAL)	FUEL USE PER CWT SHIPPED
130	10.0	0.08
400	67.6	0.17
250	25.0	0.10
		0.35
130	15.0	0.12
400	227.3	0.57
250	25.0	0.10
		0.79
		0.52
14.0	8.8	0.63
14.0		0.63
60.0	9.5	0.16
		0.16

f. Information for only one of the Co-op's three supply chains is shown. The other two are similar to Nugget Market's and are displayed in Table 12.

g. All transport in this chain is in a refrigerated box-van truck that achieves fuel economy of 10 mpg.

Source: King et al. (2010: 33).

totaling $4.3 million in 2008. Given the Co-op's high levels of community support and environmental awareness (including riding bicycles to work), it is highly likely that all the Co-op's employees live in the Sacramento area.

In Nugget's case the company is owned by a local family and seven of its nine stores are in the Sacramento area. Nugget hires approximately 1,500 employees, 60 percent of whom are full-time. Its payroll totals approximately $40 million. Additionally, Nor-Cal is owned by a local family and has 140 employees, with a payroll of approximately $6 million. It is highly likely that most of Nugget's and Nor-Cal's employees live in the Sacramento area.

Clearly, spring mix travels fewer miles in the direct and intermediated supply chains. However, fuel use results are mixed when factoring in transportation loads, demonstrating how product aggregation can provide fuel efficiency in local food chains (table 14). Fuel use per 100 pounds of product is clearly the lowest in the Co-op's local supply chain. The shorter transport distances offset the inefficiencies of transporting products in smaller loads more than do the full semi-trailer loads used in Nugget's mainstream supply chain. However, Fiddler's loads in the direct market supply chain are so small that it has the highest fuel use despite having the shortest transport distance and the highest fuel efficiency.

All three supply chains are involved in community building efforts. Eldon donates his unsold produce to food banks and serves on the board of the Davis Farmers Market. Nugget has a generous local support program that includes various agricultural organizations, and its management staff coaches entrepreneurship students from the local universities. Overall, the Co-op appears to have the most extensive community support program via its annual Holiday Meal, donating to various charities, and being involved with both food-related causes, such as educating local school children about whole grains and eating seasonally.

Key Lessons

Three general lessons emerge from these case studies of spring mix supply chains in the Sacramento area.

1. Despite the strong potential that intermediated supply chains offer conceptually, it is highly unlikely this structure will expand sales

of local spring mix. While growers have durable relationships with the local natural foods cooperative, they view the cooperative as a residual market for their excess supply because they earn higher returns from marketing their spring mix at farmers markets and through their CSA programs. Thus local spring mix growers are capturing significant premiums through their direct marketing efforts, which the Co-op, Nugget, and other retailers cannot pay when nonlocal spring mix is available at a much lower cost.

2. Related to the previous lesson is the fact that the mainstream supply chain is providing formidable competition in the spring mix market. Earthbound has been largely responsible for building the nation's spring mix market over the past 25 years. It started as a niche marketer and has now become a highly competitive nationwide supplier of an organic commodity. Unlike local growers, Earthbound manages production in two growing regions to be a highly reliable year-round supplier of organically grown leafy greens, and it gains substantial scale economies by using highly mechanized harvesting and processing technologies.

3. There are several cross-linkages between entities across the supply chains. The distributor for the Nugget Markets, Nor-Cal, is also one of the Co-op's distributors. While Terra Firma is a spring mix supplier to the Co-op, it also markets some of its produce (but not spring mix) through Nor-Cal. Fiddler's, the direct marketer, is also a spring mix supplier to the Co-op. This crossing of boundaries across the supply chains indicates that the entities involved are using entrepreneurial flexibility to take advantage of opportunities created by demand for locally produced foods.

Notes

1. Processing activities are described in greater detail in the mainstream case.
2. All information reported regarding Earthbound was obtained through secondary sources such as news articles and websites.
3. This history of Earthbound's milestones is a summary from its media kit (Earthbound Farm 2011: 2–3).
4. Based on the county agricultural commissioners' Crop Reports, which report the combined revenues for organic and conventional spring mix (Imperial County 2008; Monterey County 2008).

References

CNN Money. N.d. "100 Best Companies to Work for 2010." money.cnn.com/magazines /fortune/bestcompanies/2010/ (accessed 12/18/2011).

Davis Enterprise. 2010. "Walmart Give $467M to Charities, Including Yolo groups." May 9, C-7.

Earthbound Farm. 2011. "2011 Media Kit." ebfarm.com/sites/default/files/MediaKit2011 -Consumer_lores.pdf (accessed 12/19/2011).

Food Marketing Institute. 2009. "Supermarket Facts: Industry Overview 2008." fmi.org /facts_figs/?fuseaction=superfact (accessed 11/21/2009).

Hardesty, S., and P. Leff. 2010. "Determining Marketing Costs and Returns in Alternative Marketing Channels." *Renewable Agriculture and Food Systems* 25(1): 24–34.

Hardesty, S., and Y. Kusunose. 2009. "Growers' Compliance Costs for the Leafy Greens Marketing Agreement and Other Food Safety Programs." Small Farm Program Research Brief. sfp.ucdavis.edu/docs/leafygreens.pdf (accessed 12/19/2011).

Imperial County Agricultural Commissioner. 2008. "2008 Crop Report." co.imperial.ca .us/ag/Crop_&_Livestock_Reports/Crop_&_Livestock_Report_2008.pdf (accessed 12/18/2011).

King, Robert P., Michael S. Hand, Gigi DiGiacomo, Kate Clancy, Miguel I. Gómez, Shermain D. Hardesty, Larry Lev, and Edward W. McLaughlin. 2010. *Comparing the Structure, Size, and Performance of Local and Mainstream Food Supply Chains.* ERR-99. Washington DC: U.S. Department of Agriculture, Economic Research Service.

Local Harvest. N.d. localharvest.org (accessed 12/30/2009).

Monterey County Agricultural Commissioner. 2008. "2008 Crop Report." ag.co.monterey .ca.us/assets/resources/assets/58/crop_report_2008.pdf?1295564864 (accessed 12/18/2011).

Ramsay Highlander. N.d. ramsayhighlander.com/products/spinach-spring/spinach .htm (accessed 12/19/2011).

San Benito County Agricultural Commissioner. 2008. "2008 Crop Report." cosb.us/wp -content/uploads/2008-Crop-Report.pdf (accessed 12/18/2011).

Seltzer, Jonathan M., Jeff Rush, and Jean D. Kinsey. 2009. "Natural Selection: 2006 E. coli Recall of Fresh Spinach: A Case Study by the Food Industry Center." Food Industry Center, University of Minnesota. ageconsearch.umn.edu/handle/54784 (accessed 12/18/2011).

Stromberg, Eric. 2009. "General Manager's Report: Modern Values, Old Fashioned Nutrition." *Natural Choices,* October 2009, 2.

Supermarket News. N.d. "SN's Top 75 Retailers for 2009." supermarketnews.com/profiles /top75/2009-top-75/index2.html (accessed 12/18/2011).

U.S. Census Bureau. 2010. "The 2010 Statistical Abstract." census.gov/compendia /statab/2010/2010edition.html (accessed 12/18/2011).

USDA-AMS. N.d. "Farmers Market Search." U.S. Department of Agriculture, Agricultural Marketing Service. search.ams.usda.gov/farmersmarkets/ (accessed 12/30/2009).

———. 2011a. "Proposed NLGMA—Questions and Answers." U.S. Department of Agriculture, Agricultural Marketing Service. ams.usda.gov/AMSv1.0/getfile?dDoc Name=STELPRDC5090539 (accessed 12/18/2011).

———. 2011b. "Proposed Marketing Agreement for Leafy Greens." U.S. Department of Agriculture, Agricultural Marketing Service. ams.usda.gov/AMSv1.0/leafygreens agreement (accessed 12/19/2011).

USDA-NASS. 2007. "County Level Data." Agricultural Census, Volume 1, Chapter 2. U.S. Department of Agriculture, National Agricultural Statistics Service. agcensus.usda .gov/Publications/2007/Full_Report/Volume_1,_Chapter_2_County_Level/index.asp (accessed 12/19/2011).

Valley Fabrication. N.d. valleyfabrication.com/babyleafcutter_stainless.html (accessed 12/28/2011).

Withers, Dawn. 2009. "Updated: Earthbound Farm Adds Equity Firm as Partner." The Packer. thepacker.com/fruit-vegetable-news/updated_earthbound_farm_adds_equity _firm_as_partner_122123084.html (accessed 12/28/11).

———. 2010. "Private Label Growing for Vegetable Products." The Packer. thepacker.com /fruit-vegetable-news/shipping-profiles/california-spring-vegetables/private_label _growing_for_vegetable_products_122157004.html (accessed 12/18/2011).

6 Beef Case Studies in the Minneapolis–St. Paul–Bloomington MSA

Robert P. King, Gigi DiGiacomo, and Gerald F. Ortmann

Introduction

This series of case studies describes the distribution of beef through three supply chains in the Minneapolis–St. Paul–Bloomington Metropolitan Statistical Area (MSA):

> natural, store-brand beef sold through an upscale supermarket chain (mainstream nonlocal food supply chain),
> locally produced grass-fed beef sold direct to consumers (direct market local food supply chain), and
> locally processed, branded grass-fed beef sold in supermarkets, restaurants, and food service outlets (intermediated local food supply chain).

Products that are produced and/or processed and distributed in Minnesota and Wisconsin are defined as *local*. Supply chains that distribute local products and convey information that enables consumers to identify the products as local are considered *local food supply chains*.

The Location: Minneapolis–St. Paul–Bloomington MSA

The Minneapolis–St. Paul–Bloomington MSA (also called the Twin Cities, figure 16) is home to 3.2 million people—more than 60 percent of Minnesota's total population. The MSA consists of 13 counties (including two in Wisconsin along the Mississippi River): Anoka, Carver, Chisago, Dakota, Hennepin, Isanti, Ramsey, Scott, Sherburne, Washington, and

MINNESOTA

WISCONSIN

☐ Twin Cities metropolitan
statistical area

Fig. 16. Twin Cities MSA.

Wright in Minnesota and Pierce and St. Croix in Wisconsin. Per capita income for the Twin Cities averaged $47,696 in 2008 compared to $40,947 nationally (Bureau of Economic Analysis n.d.).

Food production, processing, and distribution play a significant role in the Twin Cities and surrounding outstate economies. Minnesota commodity sales, for example, totaled $15.8 billion in 2007, with more than 27 million acres devoted to crop and livestock production. Production is highly concentrated in grains, oilseeds, hogs, dairy, and cattle (USDA-ERS n.d.).

Three of the nation's top 20 food processing companies—General Mills, Hormel Foods, and Cargill —are headquartered in Minnesota, two in the Twin Cities. Together they generate $22.6 billion annually in food processing sales (Food Institute 2008). At the retail level, Twin Cities–based Supervalu ranks second nationally among supermarket chains. The area is also recognized for the strength of its independent retailers and smaller regional supermarket chains, such as Coborn's, Kowalski's, and Lunds/Byerly's, many of which have begun to differentiate themselves by offering locally produced food products.

Like their retail counterparts, some Minnesota farmers brand themselves through the highly successful, 20-year-old, state-sponsored "Minnesota Grown Program" as well as several regional "Buy Local" programs. This allows them to capture demand for local products when selling direct to consumers in the Twin Cities. According to the 2007 Census of Agriculture, approximately 4,300 Minnesota farmers sell direct to consumers, generating more than $34 million in sales annually (USDA-NASS 2009). This is equivalent to $8,075 per farm or $6.69 per Minnesota consumer. The Twin Cities metro has 40 farmers markets and there are about 125 community supported agriculture (CSA) farms statewide (Local Harvest n.d.).

The Product: Beef

The majority of Minnesota beef producers manage their operations part-time, with average herd sizes of 1–49 head.[1] Beef and dairy animals slaughtered for meat are moved from farm to table using a four- to six-segment supply chain (figure 17). Ownership of animals and products often changes at each segment of the supply chain.

The majority of beef cattle are finished on grain, usually in feed-lots. Increasingly, however, animals are being finished on pasture due to growth in consumer demand for grass-fed meat. Minnesota supports 115 livestock slaughter facilities—ranking it eighth nationally in number of plants (USDA-NASS 2008). Minnesota's slaughter facilities are fairly well distributed throughout the state, though the majority of plants are concentrated around the Twin Cities, to the west of the metro area, and to the south along the Iowa border. Minnesota is well positioned to handle commercial-level processing as well as smaller batches of animals for producers and independent branded beef companies looking to market through local food supply chains.

Approximately 20 percent of Minnesota's plants are federally inspected, with beef being examined for food safety, food quality, and correct labeling under Food Safety and Inspection Service guidelines. Meat that is

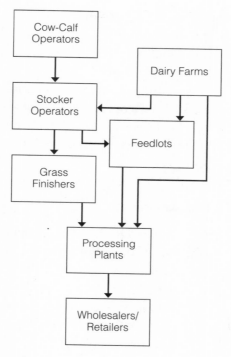

Fig. 17. Beef supply chain.

federally inspected can be sold at the retail level across state lines. Much of the beef processed at large-scale, federally inspected plants is marketed by regional and national distributors as boxed primals and subprimals through mainstream supply chains.[2]

The remaining 80 percent of livestock slaughter plants in Minnesota are registered as state "equal to" facilities operating under the State Meat and Poultry Inspection Program or as custom exempt facilities. Like meat products processed at federally inspected facilities, those processed at state "equal to" facilities can be distributed to customers through retailers, but they cannot cross state lines after processing and so may be marketed only within the same state where they are processed. Finally, custom exempt facilities are smaller still and do not allow for the resale of any product. Animals processed at custom exempt facilities must be labeled "not for sale."

After processing, Minnesota beef products are distributed to wholesalers and retailers in the Twin Cities and throughout the region. According to analysis of the Agricultural Marketing Service's *Continuing Survey of Food Intakes by Individuals*, 87 percent of all beef consumed in the United States is fresh or frozen, with ground beef accounting for almost half of all fresh beef consumed and steak accounting for another 23 percent (Davis and Lin 2005). Twin Cities beef consumption can be estimated conservatively at 208 million pounds using the national consumption rate of 65 pounds per capita.[3]

Mainstream Case: Kowalski's Markets

Kowalski's Markets is a privately held chain of nine upscale supermarkets located across the Twin Cities. Most of their stores are remodeled supermarkets purchased from independent retailers, but in 2000 they built their signature store from the ground up in Woodbury. Founded in 1983 by Jim and Mary Anne Kowalski, the company also operates a franchise Cub store, a large warehouse format store closely linked to Supervalu, the primary wholesale supplier for all the Kowalski's stores.

The focal supermarket for this case is Kowalski's Grand Avenue store. It opened in 1983 and underwent a major remodel in 1995. Located in an affluent St. Paul neighborhood near three private liberal arts colleges, this 22,000-square-foot store has average weekly sales of $425,000. It employs approximately 150 full-time and part-time workers. The meat department has seven employees—six full-time and one part-time.

Kowalski's is committed to civic engagement. The company identity statement on the homepage of their web site (kowalskis.com) asserts: "Kowalski's Companies is a civic business that organizes and governs according to democratic principles and standards. All stakeholders are responsible to renew democracy by creating a civic climate, ensuring sustainability, and growing for the common good." These values are reflected in their efforts to support local growers and to launch and foster other local businesses. Produce departments in their stores feature local fruits and vegetables whenever possible, and Kowalski's has been especially innovative in working with local producers and processors to offer fresh and processed meat products. Along with their primary line of All Natural USDA Prime and Choice beef products, they offer local grass-fed products from Thousand Hills Cattle Company (see intermediated supply chain case study).

Kowalski's is of particular interest for a local foods case study as a mainstream supermarket that is especially innovative in working with producers and processors to offer a nationally distributed natural beef product as well as a local grass-fed beef product. Kowalski's is also a company that combines a clear commitment to sustainability with a reputation for well-managed retail operations.

Supply Chain Structure, Size, and Performance

This case focuses exclusively on the supply chain for Kowalski's primary source for beef: Creekstone Farms Premium Beef. Material flows within the Kowalski's beef supply chain are depicted in figure 18.[4]

Production and Processing

Creekstone products account for nearly 95 percent of beef sales in the Grand Avenue store. Creekstone (creekstonefarms.com) was founded in 1995 by John and Carol Stewart and has facilities in Arkansas City, Kansas, and Campbellsburg, Kentucky. This family-owned business is currently organized as a limited liability company, with Sun Capital Partners, Inc. also having an ownership stake.

Creekstone offers two beef product lines under its USDA-certified branded beef program: Natural and Premium. Kowalski's purchases Prime and Choice grade Natural beef primals from Creekstone for in-store cutting and trimming. It labels these products as "Kowalski's All Natural Prime and Choice Beef."

Under Creekstone's Natural beef program, Black Angus cattle are grazed on farms throughout the central portion of the United States and finished in feedlots, where they are fed a corn-based ration with no animal byproducts. Animals in the program receive no hormones, growth promotants, or antibiotics.

According to Boyd Oase, director of Kowalski's Meat and Seafood Department, Kowalski's chose Creekstone as the sole supplier for its All Natural Beef products after a rigorous evaluation that included visits to cow-calf operations, feedlots, and the processing facilities in Arkansas

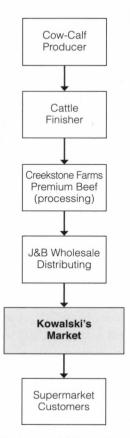

Fig. 18. Mainstream supply chain, Kowalski's Markets.

City, Kansas. Kowalski's has a verbal, long-term pricing agreement with Creekstone that is essentially a cost-plus program based on commodity prices for live animals.

Alan Verdoes, a cattle finisher in Marshall, Minnesota, is typical of those who finish independently for Creekstone. He purchases Black Angus calves from cow-calf operations in Montana, Nebraska, North Dakota, and South Dakota and finishes them to a weight of 1,300 to 1,350 pounds. Verdoes receives a premium above the conventional commodity price for beef of between $6 and $10 per hundredweight (cwt) from Creekstone (Roti 2008). The average price received by Minnesota beef finish calf operations in 2009, as reported in the Center for Farm Financial Management's FINBIN database (Center for Farm Financial Management n.d.), was $81.49/cwt. This implies an estimated price between $87.49 and $91.49/cwt for Verdoes, or a value between $1,137 and $1,189 for a 1,300-pound steer. Average feed and animal purchase costs were $967/ steer in 2009 for producers in the FINBIN database, so an estimated margin over feed and animal purchase costs for an operation like that of Verdoes would be between $170 and $222 per head.

Creekstone slaughters and processes beef at its plant in Arkansas City, Kansas. Built in 2001, the plant has a daily capacity of approximately 1,100 head and processes 250,000 head annually. The Creekstone web site notes that the plant is highly rated for animal welfare and handling. All Natural program animals are source verified, allowing Creekstone to trace beef products back to the animal's "ranch of birth" (Creekstone Farms n.d.)

Distribution

Kowalski's purchases Creekstone products through the J&B Wholesale Distributing facility in St. Michael, Minnesota. According to company information (jbgroup.com), J&B Wholesale Distributing was founded in 1979 and is a wholly owned subsidiary of J&B Group, which is the producer of No Name Premium Meats and Seafood and Midwest Pride® products. J&B has approximately 600 full-time employees—nearly all of whom live in Minnesota. It operates a fleet of more than 100 trailers serving 11 surrounding states. Cold-storage facilities are located in Pipestone and St. Michael, Minnesota; the majority of product shipped to Kowalski's comes from the St. Michael location.

J&B's distribution service is based on its commitment to product quality and food safety on behalf of clients. J&B has made substantial investments in cold chain management infrastructure (refrigerated docks, refrigerated storage, and back-up power generators) and product traceability software (UCC128). Careful cold-chain and inventory management ensures that products will not be damaged due to improper temperatures and that beef products remain within a 14 to 50-day desired window of aging and freshness. The traceability software utilizes a 128-character scan code to relay information about animal production, processing, packaging, and distribution. J&B staff members contend that most supermarket distributors do not utilize such precise traceability software. J&B also acts as an intermediary between Kowalski's and Creekstone to ensure that products meet quality specifications and food safety requirements and to address any other issues that may arise.

Despite its technological investments, J&B strives to maintain a small-company culture. J&B's ordering system relies on computers, but 90 percent of orders are still taken by a person over the phone. Orders for Creekstone boxed beef are placed through a two-tier ordering system. Oase, Kowalski's meat department director, is in direct contact weekly with Creekstone to identify and arrange product for storewide advertised specials. Once he has established a purchase plan for specials with Creekstone, he contacts J&B to place the actual order. J&B, in turn, places this order with Creekstone, incurring no risk for the "guaranteed" sale to Kowalski's. The second tier of the purchasing system involves orders placed from two to five times weekly by meat buyers to fulfill individual stores' weekly needs. This represents the bulk of beef sales. J&B estimates individual store needs and orders product accordingly from Creekstone, so J&B bears inventory risk for this portion of the supply chain. J&B sales staff members communicate by phone with Oase during a weekly conference call and with individual store meat buyers more often to discuss short- and long-term procurement plans. Steve May, territory sales manager for J&B, said in an interview for this study that his company "actively maintains a strong relationship with Kowalski's—it is the kind of relationship that we work hard at to make happen so that we can help our partners move product efficiently through the system."

Creekstone products are shipped to the J&B facility at St. Michael for warehousing and order fulfillment. Deliveries are made direct to Kowalski's stores the day after orders are placed. J&B contacts store meat buyers when product ships. Prices are based on a weekly price sheet provided by Creekstone. Kowalski's negotiates prices directly with Creekstone for all products and pays a flat per-pound overage fee to J&B for distribution. (The overage fee is confidential.) J&B requires a minimum order of 500 pounds of combined product (beef and chicken).

The distance from the J&B facility in St. Michael to the Grand Avenue store is approximately 36 miles. Deliveries are made by semi-truck with a 45,000-pound hauling capacity. The truck makes stops at several stores

Table 15. Sample beef product yield from a 1,300-pound steer

CUT	YIELD (LB)	CUT	YIELD (LB)
Ribeye	24	Roasts	64
NY Strip	20	Ribs and Brisket	26
Tenderloin	9	Stew Meat/Stir Fry/Other	42
Sirloin	24	Ground Beef	285
Total Yield 494 lb			

Source: This is based on an example in a document that describes the beef cutout calculator (National Cattlemen's Beef Association n.d.).

and typically travels a 120-mile roundtrip route. Assuming 6.0 mpg fuel efficiency, this trip requires approximately 20 gallons of diesel fuel.

Marketing and Retail Operations

The meat department in each Kowalski's store includes backroom facilities for cutting, trimming, and packaging meat from boxed beef primal and subprimals; a full-service custom meat counter; and self-service refrigerator and freezer cases. Meat and seafood account for approximately 12.5 percent of sales in a typical Kowalski's store, and beef product sales represent approximately one-third of meat and seafood sales. With $425,000 average weekly sales at the Grand Avenue store, weekly sales of meat and seafood are approximately $53,125 and weekly sales of beef products are approximately $17,700. Kowalski's All Natural beef products are labeled "Kowalski's Premium All Natural Beef. No added hormones/ antibiotics! Source verified. Product of the USA. USDA Choice/Prime." This provides the consumer with information about production methods, quality, and country of origin.

Oase notes that Kowalski's tries to price beef products so that over time the mix of cuts purchased comes close to matching the cutout from whole animals. Median prices for "Kowalski's All Natural" 85 and 87 percent lean ground beef and ribeye steak in the Grand Avenue store during 2009 were $3.99/lb and $13.99/lb, respectively.

Table 15 shows a sample whole-animal cutout for a 1,300-pound steer supplied through Creekstone. This assumes a hot carcass weight of 800 pounds and an estimated retail yield of 494 pounds of beef. With the exception of product specials, beef prices in Kowalski's stores are relatively stable—varying only in response to longer-term fluctuations in national beef prices.

For calculations in this case study, we estimate the 2009 retail value of the meat from one steer to be approximately $3,054. Assuming a 50 percent markup on cost, this implies a wholesale cost of approximately $2,037, with $750 of that going to Creekstone and J & B and the remainder to the producer/finisher.

Food Miles and Transportation Fuel Use

Meat sold in the Kowalski's Grand Avenue store typically moves through a five-segment supply chain from cow-calf producer to the retail location. Calves, weighing 600–650 pounds, are transported from Nebraska and surrounding states to farms like those owned by the Verdoes in Marshall, Minnesota, for finishing. A trip from the cow-calf operation to the Verdoes finishing lot is 250 miles one way and requires 83.3 gallons of diesel fuel for the roundtrip. Assuming that 55 live feeder cattle, which will eventually yield 27,200 pounds of meat, are hauled by semi-trailer, this implies fuel use of 0.31 gal/cwt of meat. Once ready for slaughter, live animals weighing 1,300–1,350 pounds are transported by semi-truck to Creekstone, 615 miles away in Arkansas City, Kansas, for processing. Each semi-truck of slaughter-weight steers can haul 40 animals, which will eventually yield 19,800 pounds of meat. The roundtrip requires 205 gallons of diesel fuel or 1.04 gal/cwt of meat. After processing, J & B arranges for transport of the finished meat to its distribution facility 720 miles away in St. Michael, Minnesota, and finally another 60 miles to Kowalski's Grand Avenue store using a semi-truck. With an assumed fuel efficiency of 6 mpg, total fuel use for transportation from the Creekstone plant to the J & B distribution center and then on to retail stores is 260 gallons or 0.57 gal/cwt of meat. Food miles for this chain total 1,645 miles, and total fuel use, including empty backhauls, is 1.92 gal/cwt of meat. Efficient use of transportation keeps total fuel use per cwt of product remarkably low when considering the number of miles traveled.

Community and Economic Linkages

Kowalski's is large enough to capture some economies of size in purchasing, yet small enough to have considerable flexibility in its procurement practices and store operating policies. The company is a highly respected member of the Twin Cities and Minnesota food retailing community. It has been active in the Minnesota Grocers Association as well as the National Grocers Association, and it participates in a number of community activities, including Second Harvest Heartland food bank (2harvest.org). Kowalski's does not participate actively in organizations related to local foods, nor does it link explicitly to emerging infrastructure that supports local foods activities, but the company has been active in finding ways to source local food products and features them prominently in its stores.

Finally, Kowalski's makes important contributions to the local economy by paying competitive wages to its employees and by offering convenience and excellent service to its customers. By taking advantage of economies of size and by adopting practices that maximize operating efficiency, Kowalski's is able to offer high-quality products to consumers at prices that are often lower than those in direct market channels. Much of the value-adding economic activity associated with Creekstone, however, occurs outside the region.

Prospects for Expansion

Kowalski's has grown steadily over the past quarter century, most often by purchasing and remodeling carefully selected, independently owned stores. Over the years Kowalski's stores have enjoyed solid sales growth based on excellent service and product offerings. Offering local food products has been part of the company's growth strategy. Nevertheless, local products account for only a relatively small portion of overall sales, as is the case for fresh beef.

Government policies and regulations are not viewed as a significant impediment for Kowalski's or for the trading partners in their beef supply chain. Significant investments have been made in systems that facilitate traceability, but these have been motivated more by a commitment to quality control than by regulatory requirements.

While the All Natural Beef supplied by Creekstone through J&B

captures a dominant share of its beef sales, Kowalski's also is committed to carrying Thousand Hills beef, a local product, even though its product sales account for less than 5 percent of beef sales and generate a smaller retail margin. Barring a large shift in demand for grass-fed beef or a change in relative wholesale prices for Thousand Hills and Creekstone products, sales volume shares are unlikely to change dramatically. Should there be large changes in demand or procurement prices, however, Kowalski's is prepared and able to respond quickly.

Key Lessons

Two general lessons can be drawn from this case study of a mainstream food supply chain. First, the Creekstone line of natural beef products is an example of an emerging national brand that combines sustainable production attributes with processing and distribution systems that are large enough to take advantage of significant economies of size. Highly competitive product lines like this will make it difficult for suppliers of local products (distributed through intermediated supply chains and direct market supply chains) to match quality attributes and production practices at lower prices.

Second, relatively small mainstream chains like Kowalski's have considerable flexibility in being able to offer local food products to their customers alongside products that are nationally distributed. Increases in local food product sales are most likely to be driven by improvements in upstream supply chain efficiency that make it possible to lower retail prices and by shifts in consumer demand for the attributes embodied in local products. The Thousand Hills beef supply chain, which is the focus of our intermediated case study, illustrates the feasibility of interfacing with a relatively small group of supermarkets like those operated by Kowalski's. Smaller suppliers that currently sell through direct market channels would have a more difficult challenge in being able to supply even a small retail chain like Kowalski's.

Direct Market Case: SunShineHarvest Farm

SunShineHarvest Farm (sunshineharvestfarm.com) is a small family farm located 35 miles outside the center of the Minneapolis–St.

Paul–Bloomington MSA. It is the hub of a diverse direct market supply chain that markets meat and poultry products in farmers markets, through community supported agriculture shares, and through bulk and individual item sales delivered to several drop sites. In 2008 SunShine-Harvest Farm owners Mike and Colleen Braucher marketed frozen beef from 40 animals. Beef sales were approximately $75,000 or roughly 65 percent of total gross sales.

SunShineHarvest Farm began humbly but grew quickly. In 2004 the Brauchers started direct marketing a few head of beef "to whoever wanted it." Two years later they launched their meat CSA and were among the initial vendors at the newly established Mill City Farmers Market located in the heart of downtown Minneapolis. The Brauchers expanded their beef herd and other livestock enterprises along with sales, which grew by 50 percent annually during 2006–2008.

SunShineHarvest Farm is of particular interest as a local foods case study because, like other small businesses that rely on owner-operators, its rapid growth has prompted a search for strategies to conserve time spent on marketing and deliveries while maintaining close ties with local customers who value knowing where their food comes from. Sun-ShineHarvest Farm is also a small peri-urban farm with very limited land resources.

Supply Chain Structure, Size, and Performance

Material flows within the SunShineHarvest Farm supply chain are shown in figure 19. SunShineHarvest Farm plays a central role in the chain, producing all products, setting retail prices, managing inventories and logistics, and maintaining direct personal links with all customers.

Production

SunShineHarvest Farm raises cattle, sheep, and chickens on a total of 160 acres of pasture at five locations. These include their own six-acre farm, the 40-acre farm owned by Mike Braucher's father, and three rented pastures with a combined 110 acres. Two of the three rented pasture sites are within four miles of the Braucher's homestead farm. The third, a 60-acre block, approximately 20 miles away, proved too distant to manage efficiently and Mike Braucher did not plan to use it in 2010.

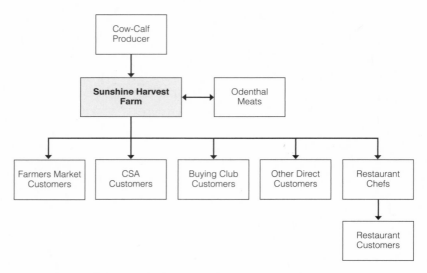

Fig. 19. Direct market supply chain, SunShineHarvest Farm.

Braucher uses management-intensive grazing practices and visits each site daily to rotate animals to a new paddock. All the pasture sites are being reconditioned through management and overseeding to improve forage value. Cows and sheep are 100 percent grass-fed. Braucher buys hay for winter feed and moves the entire cattle herd to his father's farm in winter to facilitate feeding and animal care. In late May 2009 the Brauchers had 30 cow-calf pairs, 20 yearling beef steers and heifers, 92 sheep, 500 laying hens, and 2,500 broilers. Over the course of a year they also purchase and finish 15 to 20 calves from a neighbor who feeds them exclusively on grass. In 2008 they harvested 34 steers and heifers and six cull cows.

Beef heifers and steers range in age from 24 to 30 months at slaughter. Animals are moved onto pasture in May and calving is concentrated in the spring. Cow-calf pairs graze together through October, when all animals are moved to Mike Braucher's father's farm. Braucher estimates the monthly rental rate for pasture during this period to be between $10 and $15 per cow-calf pair—approximately $80/pair for the six-month period. From November through April cows and calves are fed hay— approximately nine large bales for the pair, at a cost of $50 per bale or

$450 per pair. Yearlings go back on pasture in May. Prior to slaughter a typical steer or heifer will spend an additional nine months on pasture (at an estimated cost of $110) and six months on hay (at an estimated cost of $300), for a combined finishing cost of approximately $410. Therefore total feed costs for a typical animal raised from birth are estimated to be $940.

Processing

SunShineHarvest Farm processes three to six cattle per month at Odenthal Meats in New Prague, Minnesota (odenthalmeats.com). This family-owned and -operated facility was built in 1999 and is one of four processing plants located within 30 miles of SunShineHarvest Farm. In addition to owners Randy and Laura Odenthal, Odenthal Meats employs seven full-time and two part-time workers, all of whom live in nearby communities. The annual wage bill is approximately $210,000. This is a state-inspected plant with inspection standards "at least equal to" those imposed under the Federal Meat and Poultry Products Inspection Acts.[5] Meat from the plant can be sold directly to consumers or through grocery stores, restaurants, and food service operations in Minnesota but cannot be sold in other states. In contrast, plants under federal inspection are allowed to sell across state lines.

Odenthal Meats conducts state-inspected slaughtering one day per week—on Thursdays—and custom slaughter and processing on other days. The majority of Odenthal Meats' beef processing is done in small batches for 350 local producers under the "custom exempt" label. In a typical week Odenthal Meats processes 30 pigs and 14 cattle. SunShineHarvest Farm represents between 5 and 10 percent of Odenthal Meats' beef processing and accounts for one-fourth of all Odenthal Meats' state-inspected beef. The Odenthals also operate a meat counter at their facility where fresh beef and other meat items are sold to area customers. Approximately 90 percent of the meat sold at the counter comes from animals purchased from local producers. The Brauchers do not sell animals to Odenthal Meats.

The Brauchers use a diesel truck and trailer to deliver live animals to Odenthal Meats on Thursday mornings, when the plant does state-inspected slaughtering. The trailer can hold up to eight live cattle, but a typical load is three cattle. Beef carcasses hang in a cooler for two weeks after slaughter and are then cut up according to SunShineHarvest Farm specifications. All meat is flash frozen and vacuum packed in clear plastic or packaged in white butcher paper.

A typical steer or heifer weighs 1,100 pounds, dresses out to a hot rail weight of 600 pounds, and yields approximately 392 pounds of beef products, as shown in table 16. The price paid by consumers for meat from a whole animal ranges from an estimated $2,010 when the meat is sold in quarters and halves to buying club customers to an estimated $2,660 when the meat is sold in individual cuts at a farmers market.[6] Cull cows are processed into ground beef, with a typical yield of 325 pounds and retail value of $1,544. Netting out estimated feed costs of $940 for a steer or heifer raised from birth, then, the return over feed costs for a typical animal ranges from $1,070 to $1,720.

Odenthal Meats charges a slaughter fee of $46 per animal and $0.50/lb (hot rail weight) for cutting, wrapping, grinding, and freezing. They charge from $2.59/lb to $3.09/lb to make and package sausage or hot dogs. The processing cost for the sample steer described in table 16, which does not include any processing into sausage or hot dogs, would be $346. Netting this cost out of the return over feed costs yields a return over feed and processing costs that ranges from $754 to $1,404 per animal.

Ground beef is labeled with the Odenthal Meats logo, while all other products are packaged in clear, vacuum packaging and labeled "Braucher's SunShineHarvest Farm, 100% Grass Fed Beef, Processed for the Braucher Family, Webster MN." The Braucher's street address and telephone number as well as the "Minnesota Grown" label from the Minnesota Department of Agriculture are also included on all packaging. Consumers who purchase SunShineHarvest Farm beef know where and how the product was produced and can easily identify and contact the producers. There is full transparency and traceability.

SunShineHarvest Farm does not have a formal contract with Odenthal Meats, but the Brauchers value the feedback, advice, and service they receive through this close working relationship. They especially appreciate the flexibility Odenthal Meats offers in determining cuts of meat and package sizes. Competing processors do not allow this flexibility.

The Brauchers usually pick up beef cuts or quarters about three weeks after live animals are delivered. They haul the meat back to their farm

Table 16. Sample beef product yield from an 1,100-pound steer

CUT	YIELD (LB)	CUT	YIELD (LB)
Ribeye	16	Fajita Strips/Kabob Cubes	32
NY Strip	12	Stew Meat	16
Tenderloin	8	Ground Beef	200
Sirloin and Flat Iron	16	Oxtail	4
Roasts	60	Soup Bones	8
Ribs and Brisket	12	Liver/Heart/Tongue	8

Total Yield 392 lb

Source: SunShineHarvest Farm web site, www.sunshineharvestfarm.com.

in a freezer mounted in the bed of their diesel pickup truck. It is a 20-mile one-way trip from SunShineHarvest Farm to Odenthal Meats. The Brauchers store beef, chicken, and lamb in freezers located on their farm. They have a 441-cubic-foot stationary freezer unit, which they purchased used, that can hold more than 12,000 pounds of meat. They also store some meat in two frozen food display units they transport by trailer to farmers markets where they use them to display and store products.

Neither SunShineHarvest Farm nor Odenthal Meats has had a food safety incident for a beef product. Odenthal Meats has an approved food safety protocol in place, and they receive weekly visits from the state inspector. The Brauchers have developed simple but effective procedures for maintaining the frozen beef they sell, to ensure proper temperatures during storage, transportation, and marketing. Their labels have a Julian processing date printed on them, which allows for tracing back to the animals slaughtered on a particular date. Should some event necessitate a recall, the Brauchers could contact CSA and buying club customers fairly easily, but it would be a challenge to contact farmers market customers.

Marketing and Distribution

SunShineHarvest Farm markets beef, chicken, lamb, and eggs through a variety of channels. This enterprise mix adds complexity to their business, but they have developed an overall marketing and distribution system that offers opportunities to grow their list of customer contacts and maintain strong relationships with regular customers while economizing on the time they need to devote to marketing.

MILL CITY FARMERS MARKET AND LOCAL D'LISH: SunShine-Harvest Farm is one of three meat vendors at the Mill City Farmers Market (millcityfarmersmarket.org), and they are the only vendor for beef, chicken, and eggs. The market, which first opened in June 2006, is an important collective organization for its vendors and for the community. Located along the Mississippi River in Minneapolis between the high-traffic Mill City Museum and the Guthrie Theatre, it attracts 3,500–5,500 visitors each week. The market is open from 8:00 a.m. to 1:00 p.m. each Saturday from early May through mid-October. In 2009 all 70 vendor spaces were reserved and there was a waiting list for new producers satisfying market guidelines. Those guidelines (Mill City Farmers Market n.d.) state that all producers must carry liability insurance, and: "Priority is given to those regional farmers and producers who bring product to market that is 100% grown and harvested on farmland that they own and/or operate . . . Priority [also] is given to producers who are certified organic or use environmentally responsible growing, breeding, raising and harvesting methods." In 2008–9 the Mill City Farmers Market organized a monthly winter market at Local D'Lish (localdlish .com), a nearby store that specializes in local foods.

The Brauchers load their truck on Friday evening prior to each market day. The truck-mounted freezer is plugged into their home electricity supply overnight. They leave home at 6:30 a.m. on Saturday morning and arrive back at 3:00 p.m. Sales of all products total $1,300 for a typical week at the Mill City Farmers Market, with beef accounting for slightly more than half of the sales revenue. Contacts the Brauchers make with current and potential customers at the market are as important as the sales revenue. A purchase at the market is often the starting point for a long-term relationship that evolves into the purchase of shares in the SunShineHarvest Farm meat CSA or purchase of quarter or half animals.

Annual sales for all SunShineHarvest Farm products at the Mill City Farmers Market and Local D'Lish totaled $32,900 in 2008–approximately 29 percent of total annual sales. The Brauchers attribute approximately 53 percent, or $17,400, of this revenue to beef products. Participation in these markets required 30 trips to Minneapolis—each of which necessitated approximately 70 miles and a combined 16 hours of the Brauchers' time. Annual market membership, booth, and service fees are $748. Setting the cost of vehicle operation at $0.637/ mile (Barnes and Langworthy 2003) and the opportunity cost of the Brauchers' labor at $18.83/hour (El-Osta and Ahearn 1996), the total annual cost for their market participation is $10,378—approximately 32 percent of total sales through this market channel. If the retail value of a steer or heifer sold at the Mill City Farmers Market is $2,660, estimated marketing costs will be $851. Netting this cost out of the return over feed and processing costs yields a return over feed, processing, and marketing costs of $553 per animal for sales in this market channel.

MEAT CSA, BUYING CLUBS, AND OTHER DIRECT SALES: Sun-ShineHarvest Farm also sells direct to consumers through a CSA, a buying club, and orders placed by phone or email. The Brauchers travel to Twin Cities drop sites on three Thursdays each month to make deliveries to customers who purchase products through these channels.

CSA customers purchase a whole or half share for either three or six months. A full share is approximately 20 pounds of meat and eggs delivered once each month. A typical share includes two whole frozen chickens, two dozen eggs, four pounds of ground beef, and six pounds of other meats (roasts, steaks, stew meat, bacon, sausage, ham, etc.). Customers pay in advance, and prices range from $600 for a six-month full share to $155 for a three-month half share. In 2008, CSA sales totaled $36,700, or 32 percent of SunShineHarvest Farm's gross annual sales. By weight, beef accounts for 40 to 50 percent of a CSA share for a typical month. The Brauchers note that their customers like the simplicity and variety of the CSA. CSA sales, in turn, make it easier for the Brauchers to market the whole animal and greatly reduce the time required for preparing orders and collecting payment. Also, because no cash needs to be collected, they typically need only 15 minutes to deliver shares to a CSA drop site. We estimate the whole animal value for beef marketed to CSA customers to be $2,010.

On one of their monthly Thursday evening trips to the Twin Cities, the Brauchers deliver orders to members of a self-organized buying club.

Organized by a local foods advocate who learned about SunShineHarvest Farm on the Internet, the club has about 30 members, most of whom are linked through participation in home schooling or church membership. Buying club members can pre-order individual cuts of meat at farmers market prices, and they receive a small discount on ground beef and eggs when ordering in bulk. Money is collected from buying club members at the time of delivery. In 2008 buying club sales totaled $18,000, or approximately 16 percent of SunShineHarvest Farm total annual sales. The Brauchers attribute approximately 26 percent, or $4,600, of this revenue to beef products. Preparation of buying club orders is more time-consuming than assembling CSA shares, and buying club deliveries take much more time—typically more than one hour—because it is necessary to collect payment. Nevertheless, this is a valued market outlet for SunShineHarvest Farm. We estimate the whole animal value for beef sold to buying clubs to be $2,470 for individual cuts of meat and $2,010 for halves and quarters.

Customers also can contact SunShineHarvest Farm directly to purchase standard boxes of beef products or a half or quarter beef. The standard beef boxes include a 25-pound variety box priced at $160, a 10-pound sample box priced at $65, and a 10-pound ground beef box priced at $50. Quarter and half beef purchases cut to customer specifications are priced at $3.00 per pound hanging carcass weight plus a processing cost of approximately $0.60 per pound. A typical quarter of beef has a hanging weight of 150 pounds, costs $540, and yields approximately 98 pounds of frozen beef. Therefore revenue from a whole animal marketed in this way is $2,160. Orders for halves and quarters can be picked up at a drop site or at the Mill City Farmers Market. Sales through this channel totaled approximately $18,000 in 2008–16 percent of total sales and 24 percent of beef sales.

SunShineHarvest Farm also sells a very small amount of beef ($1,000, or 1 percent of all beef sales in 2008) to several restaurants, including Spoonriver in Minneapolis. Spoonriver owner and chef Brenda Langton purchases primals and offers "SunShineHarvest Grassfed Beef" as a daily special—changing the entrée as needed in an effort to utilize the whole animal. Langton began ordering SunShineHarvest Farm beef exclusively in 2009 and increased her purchases to $450/week. She says her customers value the SunShineHarvest Farm beef quality and taste, which she believes is due to the grass-fed production standards and the dry-aging process employed by Odenthal Meats. Deliveries to Spoonriver and other restaurants are usually made on a Thursday evening trip to the Twin Cities or on a Saturday trip to the Mill City Farmers Market. We estimate the whole animal value for beef sold in this channel to be $2,470.

A typical Thursday evening drop site delivery requires approximately 70 miles of driving roundtrip and four hours of Mike's time for order preparation and delivery. Once again setting the cost of vehicle operation at $0.637 per mile and the opportunity cost of the Brauchers' labor at $18.83 per hour, the total cost for 36 trips over the course of a year is $4,317. With annual sales through the direct purchase channels served by these deliveries totaling $80,900, the delivery cost represents only about 5 percent of the associated sales revenue. We estimate the average whole animal value for the beef sold to CSA, buying club, and restaurant customers to be $2,100. This implies marketing costs of only $105 per animal. Netting out these costs along with feed and processing costs yields a return of $739 per animal. Even though gross revenue per animal is lower in this channel, the net return to SunShineHarvest Farm is higher than that for beef sold in farmers markets. Returns for both direct market channels are well above the $45 return over direct expenses (feed, transportation, and marketing) received by beef finishers in 2009 who sold into commodity markets, as reported in the FINBIN database (Center for Farm Financial Management n.d.). It is important to note, however, that the market for direct marketed beef is small and that large price reductions may be required to balance demand and supply if there are significant increases in supply.

Food Miles and Transportation Fuel Use

The SunShineHarvest Farm supply chain involves transportation of live animals from the farm in Webster to the Odenthal Meats facility in New Prague, and then a return haul of the frozen meat products. Combined, these two 40-mile roundtrips in a diesel pickup truck with a fuel efficiency of 16 mpg require 5.0 gallons of diesel fuel. Assuming three animals are transported on a typical trip, with an associated meat yield of 1,176 pounds, this implies fuel use of 0.42 gal/cwt of meat.

The downstream portion of the supply chain involves transportation of frozen meat products to the Mill City Farmers Market or a drop site in the Twin Cities metro area. On average this is a 70-mile roundtrip in the diesel pickup that requires 4.4 gallons of diesel fuel. Assuming 250 pounds of product are sold on a typical trip, this implies fuel use of 1.76 gal/cwt of meat. Adding pre- and post-processing fuel use together yields a total of 2.18 gal/cwt of meat. Total miles traveled by a steer or heifer from the farm to the customer averages 75 miles—20 miles from the farm to the processor, 20 miles from the processor back to the farm, and 35 miles to a market or drop site in the Twin Cities.

Community and Economic Linkages

SunShineHarvest Farm does not participate in mainstream beef organizations such as the Minnesota State Cattlemen's Association, nor does it have linkages to the processing and distribution segments of the commodity beef sector in the state or region. In contrast, SunShineHarvest Farm does benefit significantly from linkages to the strong local foods movement in the Twin Cities metropolitan area. As a vendor at the popular Mill City Farmers Market, the farm gains valuable access to customers and visibility. In July 2009 SunShineHarvest Farm was featured in the Taste section of the *Star Tribune*, a major daily newspaper with metro-wide distribution (Moran 2009). SunShineHarvest Farm also reaches potential customers through listings in the Minnesota Grown Directory (www3.mda.state.mn.us/mngrown) and on web sites such as Local Harvest (localharvest.org/farms/M11356), and Eat Wild (eatwild.com /products/minnesota.html).

The Brauchers are active members of the Cannon River–Hiawatha Valley Chapter of the Sustainable Farming Association of Minnesota (sfa-mn .org). This organization "supports the development and enhancement of sustainable farming systems through farmer-to-farmer networking, innovation, demonstration, and education" (Sustainable Farming Association of Minnesota n.d.). The Brauchers also contribute to and benefit from the community of vendors at the Mill City Farmers Market.

The Brauchers' Thursday evening drop site visits establish small communities of customers who interact when they pick up food. This applies especially to the buying club, which was established by a customer as a way to give others access to SunShineHarvest Farm product. The Brauchers occasionally host farm visits organized by the Mill City Farmers Market. They also welcome their regular customers to visit the farm and sometimes invite prospective CSA customers to see their operation.

Finally, SunShineHarvest Farm is currently a part-time business for the Brauchers. Colleen Braucher has a full-time job off the farm. Mike Braucher is transitioning to full-time work on the farm, but he still works part-time as an engineer. In part because Mike Braucher is giving up relatively high-wage work, the overall impact of SunShineHarvest Farm on the local economy is relatively small and may even be negative. SunShineHarvest Farm is an important customer for Odenthal Meats but represents less than 10 percent of the processor's business. If SunShineHarvest Farm were to cease operations, Odenthal Meats could likely recover some of the lost revenue by attracting new processing customers who want to take advantage of their state-inspected status. On the positive side, Odenthal Meats has some unused capacity that would allow it to handle a 50 percent increase in volume from SunShineHarvest Farm without major disruption to its business with other processing customers. Here again, the marginal impact of SunShineHarvest Farm's operations on the local economy is probably relatively small.

Prospects for Expansion

SunShineHarvest Farm has grown rapidly, and prospects for continued growth are good. The Brauchers are remarkably effective direct marketers. They are approachable, engaging, and good listeners. One need only watch them in action for a few minutes at the Mill City market, for example, to see they enjoy what they are doing and easily establish comfortable relationships with their customers.

Mike Braucher is scaling back on his off-farm work. This will give him more time to devote to both production and marketing activities, but Colleen will retain her off-farm job as a source of income and benefits. On the production side, access to land appears to be the most significant immediate barrier to expansion. Having animals on pasture in five separate locations adds significantly to the time required for animal care and pasture management. Though SunShineHarvest Farm has stable leases on its three rented sites, uncertainty about having long-term access is a cause for concern. In addition, with such limited land resources, SunShineHarvest Farm must purchase all the hay needed for winter feed. This puts them at a potential disadvantage relative to producers who can pasture animals and produce hay on their own land. The Brauchers have been exploring opportunities to purchase a larger farm. They want to stay close to their markets in the Twin Cities metro area. However, land values are very high in their peri-urban county, and prices do not decline appreciably in slightly more distant counties. Also, their capital resources are limited, and the weak housing market makes it difficult to sell their current six-acre farm and home site.

The Brauchers have not felt unduly hindered by government policies and regulations. Though not required to, Mike Braucher has a food handler's license. He also has a food preparation certificate that allows him to prepare food for samples at farmers markets. Looking to the future, the Brauchers, like other small livestock producers, express concerns about proposed national animal identification programs. They are unsure about what the costs would be and about what impacts such programs could have on their operation if implemented.

Despite strong demand for local foods and grass-fed beef, SunShineHarvest Farm also faces challenges on the marketing side. Their products are priced well above mainstream retail prices. This can be a barrier to demand growth as consumers scale back in response to difficult economic conditions. There is also increased competition from local branded products such as the grass-fed beef marketed by Thousand Hills Cattle Company (see intermediated supply chain case study). Thousand Hills sells fresh grass-fed beef in supermarkets and food cooperatives around the Twin Cities metro area. They too have been growing rapidly, and their presence in high traffic stores makes it easier to attract new customers. SunShineHarvest Farm also faces strong competition from several other direct marketers who offer grass-fed beef products at farmers markets or through CSAs. While this competition imposes some limits on opportunities for SunShineHarvest Farm, the Brauchers also believe that having a community of farms and businesses that share their dedication to quality and sustainable production practices helps develop the market for everyone.

Key Lessons

Three general lessons about direct marketing supply chains emerge from this case study. First, it exemplifies the importance of personal relationships between producers who direct market and their customers. SunShineHarvest Farm commands a price premium for its products. It delivers high quality products but not necessarily the product consistency one would expect in a branded product sold through mainstream distribution channels. SunShineHarvest Farm customers realize that there will be some variation in product quality and, for CSA customers, product mix. The Brauchers are the "constant" in the relationship. This has been the basis for sales growth, and it has given SunShineHarvest Farm greater flexibility in meeting the challenge of marketing the "whole" animal in a cost-effective manner. The multi-pronged distribution strategy the Brauchers have developed offers opportunities for them to interact with large numbers of potential customers in the Mill City Farmers Market and allows loyal customers to migrate to distribution channels with lower prices for customers and lower transaction costs for SunShineHarvest Farm. This is a strategy that can be and is being adopted by other farm operations that market directly to consumers.

Second, reliance on personal relationships also creates challenges for scaling an operation up to the point where it can sustainably support at least one person full-time. During the months that the Mill City Farmers Market is in operation, the Brauchers devote an estimated 26 hours per week to marketing and delivery activities that bring them into direct contact with their customers, and they are considering the addition of another farmers market to their already busy schedule. As they add new customers, they can realize some size efficiencies in their weekly deliveries, but at some point this growth may necessitate adding another weekday evening delivery. These marketing and distribution challenges in scaling up are common to farm operations that market directly to consumers.

Finally, markets for locally produced, differentiated food products like the grass-fed beef sold by SunShineHarvest Farm are growing, but they are also small and thin. As new operations enter direct-to-consumer markets, there is likely to be downward pressure on prices as more sellers compete for a relatively small number of customers. Building strong long-term relationships with customers can help insulate a farm from this downward pressure on prices, but charging higher prices may make it difficult to offset natural attrition of long-term customers. Challenges from branded products offered in mainstream supermarkets may be still more significant for direct market operations like SunShineHarvest Farm. These products often are competitively priced and may offer more consistent product quality and greater convenience for consumers.

Intermediated Case: Thousand Hills Cattle Company

Thousand Hills Cattle Company is a privately held business that markets "gourmet quality" grass-fed beef in the Twin Cities metro area. It is the lead entity within an intermediated supply chain. At the upstream end, the close, long-term relationships Thousand Hills has cultivated with its producers and processor, the scale of operations and mode of distribution, and the unique attributes of its products all are distinctly different from the mainstream supply chain for grain-finished beef. At the downstream end, however, the Thousand Hills products reach consumers through mainstream supermarkets, high-end restaurants, and institutional food service operations.

Founded in 2003 by Todd Churchill, Thousand Hills has grown rapidly and currently markets meat from 1,300 cattle annually out of its 10,000-square-foot facility in Cannon Falls, Minnesota. Churchill grew up on a farm in Illinois and graduated with a business degree from St. Olaf College in nearby Northfield. He is a CPA and owned an accounting practice prior to founding Thousand Hills. One of his clients was Lorentz Meats, the Cannon Falls–based company that processes meat for Thousand Hills and now is a key partner in the Thousand Hills supply chain.

The headline on the Thousand Hills web site (thousandhillscattleco .com) reads: "Our 100% grass fed beef is not only delicious, but good for your health and locally produced." This statement summarizes the value proposition that is the basis for their marketing strategy. First and foremost, Thousand Hills emphasizes that their beef has outstanding taste and quality. Also important, but secondary to taste and quality, Thousand Hills publicizes the health benefits of grass-fed beef and the humane, sustainable farming practices of the family farmers from Minnesota and surrounding states who raise cattle for Thousand Hills.

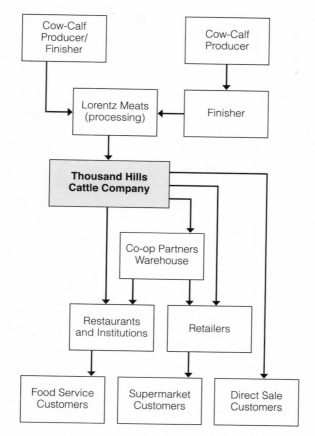

Fig. 20. Intermediated supply chain, Thousand Hills Cattle Company.

Supply Chain Structure, Size, and Performance

Material flows within the Thousand Hills supply chain are shown in figure 20. Thousand Hills coordinates the chain, establishing protocols for beef production and processing, setting prices for cattle procurement and wholesale meat sales, and managing inventories and logistics.

Production

Approximately 40 producers located in Minnesota, Wisconsin, Iowa, Nebraska, North Dakota, and South Dakota raise cattle for Thousand Hills. All conform to a strict production protocol that specifies allowable feeding, husbandry, veterinary care, source verification practices, and terms of sale (Thousand Hills Cattle Company n.d.). Some producers have beef cows and finish calves. Others source calves from elsewhere, typically from the Dakotas and Nebraska, and finish them closer to Cannon Falls. Churchill works closely with all producers and visits each farm several times each year. He does not actively recruit producers. Rather, he prefers that they approach him after they have made a decision to convert to grass-fed production.

Churchill finishes approximately 11 percent of the Thousand Hills cattle supply—140 head—on his own land located near Cannon Falls. He purchases most of these animals as calves or yearlings but is also experimenting with cow-calf production. These experiences on his own farm help Churchill refine the production protocol for Thousand Hills.

Churchill assesses animal quality during his farm visits. Based on visual inspections, he can usually tell producers which animals will meet Thousand Hills' standards by the time the animals weigh 500 to 600 pounds. This gives producers time to find other market outlets for those cattle that Thousand Hills will not purchase. The farm visits also help Churchill estimate finish dates and schedule cattle deliveries, since Thousand Hills processes at a steady rate year-round. Producers do not have preestablished delivery windows. Rather, Churchill and producers track rates of weight gain and work together to plan delivery dates. Churchill's own cattle are a buffer inventory of live animals that help him fill processing and delivery gaps.

In 2009 producers received a base price of \$1.75/lb hot carcass weight, but cattle grading out as standard or "low select" were discounted by \$0.25/lb. A steer that weighs 1,200 pounds live will normally have a hot carcass weight of approximately 650 pounds, according to Churchill, and will yield an undiscounted pay price to the producer of \$1,138 from Thousand Hills plus a \$20/head allowance for transportation costs. The producer pays any remaining transportation cost. Producers receive final payment from Thousand Hills within 10 business days of delivery.

Jim Larsen has supplied finished beef to Thousand Hills since the company began operations and is typical of Thousand Hills producers.[7] He markets 25 purebred Black Angus steers and heifers in a typical year, selling about half of these animals to Thousand Hills. The rest are processed as halves and quarters to direct market customers. He also does custom transport work for Thousand Hills and conducts in-store demos. In the past, he worked in the Thousand Hills warehouse.

Larsen, who is in his 70s, operates a 200-acre farm, located 15 miles west of Cannon Falls, that has been in his family since 1866. His wife is employed off the farm. Larsen worked in banking and farm equipment and feed sales in Montana and Nebraska for 35 years before returning to the farm in 1992. He gradually transitioned the farm (while taking over from his father) from corn and soybeans to grass for pasture and hay, seeding pastures for the first time in 1997. He currently has 45 cow-calf pairs, and he purchased an additional eight steers in 2009 for finishing.

Larsen follows the Thousand Hills production protocol for all the animals he produces. Cows and yearlings go on pasture in mid-April, and calving is concentrated as close to May 1 as possible. Larsen's cows have few calving problems, and he normally uses holistic and herbal remedies to address the few health issues that do arise. Animals are fed on pasture until November. Larsen reduces his herd by half before winter by selling bred cows to other farmers interested in starting grass-fed beef production. This reduces his need for hay over the winter.

Larsen keeps two bulls with the cows, and the insemination rate is nearly 100 percent. He uses rotational grazing, though he also makes hay available year-round because he believes animals like to balance grass feeding with some dry hay. He has divided his pasture into five-acre paddocks. The herd is divided into three groups, and each group moves to a new paddock every two or three days. Larsen fertilizes his pastures using liquid fish emulsion and composted manure. The goal is to raise the sugar level in the grass in order to increase daily gain. He estimates his fertilizer cost is $20 to $25 per acre annually. Animals are fed hay through the winter. Larsen raises hay on 60 of his own acres and on an additional 25 rented acres. He also purchases 100–125 tons of hay per year at a cost of $8,000–$10,000. Calves are weaned at 10 months and are ready to market at a live weight of 1,200–1,300 pounds as early as at 18 months.

Setting hay costs at $9,000 and adding to this a rental rate/opportunity cost of land assumed to be $22,500 per year (225 acres at $100/acre) and fertilizer costs estimated to be $3,150 (140 acres at $22.50/acre), Larsen's feed costs are approximately $34,650 per year. Not all of this cost can be attributed to beef finishing, since he sells some bred heifers, bull calves, and cull cows. If beef finishing accounts for two-thirds of the total cost, Larsen's feed cost per head of finished beef is between $900 and $1,000. If a 1,200-pound steer has a hot carcass weight of 650 pounds, Larsen will receive $1,138 from Thousand Hills. He receives $2.25/pound plus processing costs for the halves and quarters he sells to direct market customers, so the revenue for the same steer sold direct would be $1,463. Therefore Larsen's margin over feed costs is between $138/head and $563/head, which is well above the $45 return over direct expenses received in 2009 by beef finishers who sold into commodity markets (Center for Farm Financial Management n.d.).

Processing

Thousand Hills delivers 25 cattle weekly to Lorentz Meats—also located in Cannon Falls—on Thursday evening or Friday morning. The cattle are slaughtered on Friday morning before any other animals are slaughtered. Carcasses hang for five days and then are cut and packaged according to Thousand Hills specifications, which vary greatly from week to week due to fluctuations in demand.

Lorentz Meats (lorentzmeats.com) operates a 10,000-square-foot plant equipped to process and vacuum package fresh and frozen primal cuts as well as retail, case-ready products. They can grind, smoke, and cook meat to specification and produce a variety of processed beef products such as hot dogs and meatballs with wild rice. The company, organized as a C-corporation, is managed by part-owners Rob and Mike Lorentz, who purchased the 30-year-old family business from their parents in 1997. In 2001 the Lorentz brothers replaced their aging facility with a new, state-of-the art processing plant. The new plant, which is USDA inspected and certified organic, was designed to handle small batch processing required by direct marketers and companies such as Thousand Hills that sell into specialty niche markets. Lorentz Meats is recognized nationally for its humane treatment of animals, high standards for product quality, and willingness to work with direct market producers.

Lorentz Meats processes cattle, bison, and hogs five days each week. Their customers include 300 small farmers and three large accounts that process significant numbers of animals on a regular basis year-round. Thousand Hills is one of these three large clients and accounts for approximately 12 percent of Lorentz Meats' business. Daily capacity is 40 beef or bison or 120 hogs. In addition to the Lorentz brothers, the plant employs 54 staff. Six of these are salaried, and the remaining 48 are paid on an hourly basis. Seven employees work on a second shift that is exclusively dedicated to packaging. The annual wage and salary bill for Lorentz Meats is approximately $1,440,000. Employees also have the option to purchase health insurance through the company and can participate in an optional retirement plan with some company match to employee contributions. The plant pays more than $30,000 annually in property taxes. Therefore its operations contribute significantly to the local economy, not only by adding considerable value to local agricultural production but also through the wages and taxes it pays.

Processing fees at Lorentz Meats for direct market producers are $0.58 per pound hot carcass weight for cutting and packaging plus a splitting fee. In 2009 they also required a $28 per animal rendering surcharge due to the depressed market for animal byproducts. The per-animal cost for processing varies greatly, but a typical cost is approximately $400. Mike Lorentz declined to provide a schedule of processing fees for Thousand Hills. He reported that their base processing cost per pound is significantly discounted. The discount received by Thousand Hills is a function of volume and, more important, reduced cutting and packaging associated with primal cuts. There are additional fees for further cutting, processing, and packaging. These vary significantly from week to week, as do cutout specifications. When a Thousand Hills animal is processed completely to retail case-ready cuts—which require more trimming and greater quality control than comparable cuts for direct market customers—Mike Lorentz estimates that the cost approaches the $400 per animal fee paid by direct market customers. Finally, Mike Lorentz notes that Thousand Hills sometimes uses a third party for some further processing, especially ground beef production. No further information was available on costs or product volumes for this third party processing.

Lorentz Meats records detailed information on each animal processed. They provide weekly information to Thousand Hills on a batch basis, though they can provide cutout information on individual animals if so requested. Thousand Hills does receive the hot carcass weight for each animal, since that is the basis for payment to producers. Thousand Hills owns the cattle when they are processed, therefore Lorentz Meats sends all production information to Churchill rather than to individual producers.

Lorentz Meats plays a critical role in assuring product quality and food safety for Thousand Hills. Lorentz Meats has an excellent food safety record, and they have processing procedures and testing equipment that conform to the quality and safety standards of Thousand Hills customers. When asked about food safety, Churchill noted that Thousand Hills addresses food safety concerns, such as *E. coli*, through management and feeding programs that strictly limit animal contact with fecal matter. "Cattle coming from our pastures do not have manure on hide," Churchill said. "They are clean going onto the kill floor." He went on to say that no Thousand Hills animals have been condemned or cited for a food safety incident.

The relationship between Thousand Hills and Lorentz Meats is characterized by mutual trust and respect. Each business contributes significantly to the success of the other. The knowledge and business skills embodied in these two firms are highly complementary, and in combination they help ensure a very high quality product. Finally, it is noteworthy that both Thousand Hills and Lorentz Meats are linked to other local food supply chains. Thousand Hills is beginning to provide distribution services to local poultry producers and to Lorentz Meats for their processed pork products. Lorentz Meats, in addition to processing for many direct market producers, has been active in providing educational programs for direct market livestock producers around the region and nationally.

Marketing and Distribution

Thousand Hills has enough cold storage and freezer space in their Cannon Falls facility to hold the meat from 40 to 50 animals. Fresh ground beef has a 30-day shelf life, while the shelf life for other fresh products is 40 to 60 days. Churchill's goal is to sell everything fresh, but a small amount of meat is frozen and sold direct to consumers who place orders via phone or the Internet.

In addition to working with farmers, Churchill is responsible for new account development and is the public face of the company. Thousand

Hills employs four full-time and two part-time staff who run day-to-day operations related to marketing and distribution. They handle orders, delivery schedules, and customer service. Thousand Hills also employs several part-time demonstration staff, some of whom are farmers. Their interactions with customers are a valuable source of feedback and market information.

The lead time on sales is three to four days. Customers typically place orders on Monday, and most deliveries are made on Thursday and Friday. Thousand Hills owns a 16-foot refrigerated delivery truck with a 10,000-pound capacity. This truck makes weekly direct deliveries to retail grocery stores, restaurants, and institutional customers. The only exception to this direct-delivery model is the use of Coop Partners Warehouse (coopartners.coop) in St. Paul, Minnesota, for some orders placed by natural food cooperatives, but the bulk of product volume reaches retail outlets by direct delivery.

Most retail outlets order product in case-ready packaging, though some customers order boxed beef in quantities that may or may not balance out to whole carcasses. Restaurants order either standard retail cuts or primals. Most institutional sales are either bulk ground beef or processed products, such as wieners.

In July 2009 the Thousand Hills web site listed 60 grocery stores in Minnesota, two in Wisconsin, one in Iowa, and one in Illinois that carry its products. Thousand Hills products are regular menu items for two Twin Cities restaurants and are listed as "specials" on menus for several others. Institutional customers include St. Olaf College and several school districts that have featured Thousand Hills products in farm-to-school programs. By design, the customer base is highly diversified. No single retail or food service establishment represents more than 4 percent of total sales.

Thousand Hills has a standard wholesale price list and provides suggested retail prices. Typical retail margins are approximately one-fourth of the retail price—a 33 percent markup on wholesale cost—but margins vary across stores. Thousand Hills works with retailers on promotional strategies and typically bears the costs for promotions.

Thousand Hills does not have a typical cutout for its animals. Non-promotion retail prices vary considerably across retail locations, by as much as $2.00/lb for ground beef ($4.99 to $6.99/lb) and $5.00/lb for

Table 17. Sample beef product yield from a 1,200-pound steer

CUT	YIELD (LB)	CUT	YIELD (LB)
Tenderloin Fillet	7	Chuck Arm Roast	42
Boneless Ribeye Steak	22	Round Rump Roast	14
New York Strip Steak	20	Stew Meat	16
Top Sirloin Steak	10	Stir Fry Meat	16
Round Steak	24	Skinless Hot Dogs	18
Ball Tip Steak	11	Ground Beef	224
Total Yield 424 lb			

Source: Thousand Hill Cattle Company web site, www.thousandhillscattleco.com.

ribeye steak ($16.99 to $21.99/lb) in the same week at the six Twin Cities locations monitored for this study. Median prices across all market locations in 2009 were $5.99/lb for ground beef and $17.99/lb for ribeye steak.

To estimate the retail value of a typical animal, we used cutout and price information for a 53-pound "Value Pak" advertised on the Thousand Hills web site as equivalent to a 1/8 carcass. Quantities for individual cuts and processed products in the Value Pak were multiplied by eight to determine the "typical" cutout information presented in table 17, and the $380 Value Pak price was multiplied by eight to yield an estimated retail value for a whole animal of $3,040.[8] This implies an average price of $7.17/lb across all cuts for Thousand Hills beef.

Retail Sales

Kowalski's is typical of the retail firms that carry Thousand Hills products.[9] They operate nine upscale supermarkets located around the Twin Cities metropolitan area. In addition to Thousand Hills products, they sell prime and choice natural beef sourced from Creekstone Farms in Arkansas City, Kansas. A typical Kowalski's store orders a mix of primals and case-ready cuts twice weekly from Thousand Hills and receives delivery to the store the day after orders are placed. Thousand Hills products are displayed next to the natural beef products in refrigerated cases and in the custom meat counter. For meat cut from primals, Thousand Hills provides labels that are affixed on store packages. Kowalski's emphasizes quality and service in its meat departments and was among the first Twin Cities retailers to carry Thousand Hills products. Although Thousand Hills products account for less than 5 percent of beef sales and yield a lower margin than the natural beef, Kowalski's values the opportunity to give its customers a chance to purchase this locally produced meat.

All Thousand Hills consumer-ready products are labeled: "100% grass-fed beef. Pasture raised on local farms and source verified. Not given hormones, antibiotics, or animal by-products. U.S. inspected and passed by the Department of Agriculture. Cannon Falls MN." This provides the consumer with information about production methods, product quality, and processing location. Particular reference is made to local farms, but individual producers are not identified.

Food Miles and Transportation Fuel Use

In a representative case, Thousand Hills beef is moved approximately 300 miles from farm to market using a five-segment supply chain as already mentioned. Live animals at slaughter weight are typically moved 250 miles from farm to processor in a small semi-truck that has a fuel efficiency rate of 9 mpg, requiring almost 56 gallons of diesel fuel for the 500-mile roundtrip. We assume 27 animals are moved on each trip, with an average live weight of 1,100 pounds and an average meat yield of 424 pounds per animal. This implies a total beef yield of 11,448 pounds. Therefore, total fuel use for this trip would be 0.49 gal/cwt of beef. The downstream portion of the supply chain involves moving fresh, finished beef products from Lorentz Meats to the Thousand Hills warehouse and then direct to retailers in the Twin Cities or to Coop Partners Warehouse. The short 10-mile roundtrip from Lorentz Meats to the warehouse requires 2.0 gallons of fuel to haul 10,600 pounds of meat, or 0.02 gal/cwt of beef. On average the 90-mile roundtrip to the Twin Cities uses 14.0 gallons of fuel or 0.18 gal/cwt of beef transported in a Thousand Hills 16-foot refrigerated truck with a 6.5 mpg fuel efficiency rate and a hauling capacity of 9,500 pounds when 80 percent full (Thousand Hills does not typically fill its truck when moving product to the Twin Cities).

Overall, a steer travels a total of 300 food miles from farm to consumer. Total truck fuel use is 0.69 gal/cwt of beef.

When animals are born and raised on Jim Larsen's farm, just 15 miles west of Cannon Falls, total food miles fall to 62 miles, and truck fuel use falls to 0.23 gal/cwt of beef. Conversely, when animals are bred on cow-calf operations in Nebraska and finished in Minnesota, as is the case for some of the Thousand Hills beef, total food miles rise to 607 miles and truck fuel use increases to 0.98 gal/cwt of beef.

Community and Economic Linkages

Thousand Hills does not participate in mainstream state beef organizations such as the Minnesota State Cattlemen's Association and Minnesota Beef Council. However, Thousand Hills does organize an annual meeting for its beef producers in the autumn at one of Churchill's own production sites. This offers an opportunity for educational programs on beef production practices, pasture management, and meat quality. It also gives producers a chance to meet and exchange ideas. Representatives from retail outlets and restaurants are sometimes invited to these events.

Thousand Hills is a member of the Land Stewardship Project (land stewardshipproject.org), a nonprofit organization founded in 1982 to promote sustainable agriculture through educational programming and policy advocacy. Churchill is often invited to speak to groups interested in sustainable agriculture and local food systems. His speeches focus on claims regarding the quality and health benefits of grass-fed beef and on its environmental and food safety advantages relative to mainstream feedlot production.

Thousand Hills also participates in and benefits from both the mainstream and local food system infrastructure that is not necessarily linked to beef. The Thousand Hills supply chain links into the mainstream food system of the Twin Cities metro area at the retail level—both through supermarkets and upscale restaurants. This enables Thousand Hills to reach large numbers of customers at a low cost. Also, as noted earlier, Thousand Hills uses Coop Partners Warehouse for some deliveries to natural food cooperatives. In this way the company benefits from collectively organized local food distribution infrastructure originally developed for other products.

In an effort to use its own distribution resources more efficiently, Thousand Hills is also contributing to the local food infrastructure. Thousand Hills uses its trucks to transport Lorentz Meats' processed pork products to retail outlets and is exploring marketing and distribution partnerships with local free-range poultry operations.

Thousand Hills has four full-time and two part-time employees as well as six to seven part-time people who do product demonstrations. The estimated annual wage bill is $210,000, not including proprietor income for Churchill. If Thousand Hills accounts for approximately 12 percent of the wage bill for Lorentz Meats, this adds another $172,800 in wage income, as well as proprietor income for Mike and Rob Lorentz, to the economy of the region surrounding Cannon Falls. Both businesses also pay significant property taxes, and the premium prices paid to farmers by Thousand Hills add to the income of beef producers in the region. Therefore the Thousand Hills supply chain has a significant positive impact on the regional economy.

Prospects for Expansion

According to Churchill, the target market for Thousand Hills is consumers who want to purchase individual cuts of fresh meat in places where they normally do a significant portion of their grocery shopping. Churchill goes on to say that his customers are interested in buying local, but they are not interested in buying direct if it means "giving up on convenience" nor in purchasing large quantities (such as meat from a half or whole animal) in a single transaction. Consumers pay attention to signals provided by their retailer that the foods they are buying are locally and/or sustainably produced. Churchill believes these same consumers appreciate seeing Thousand Hills listed as a supplier on restaurant menus, and they value having Thousand Hills products offered to their children in a school or college food service program.

Thousand Hills sales have been growing at an annual rate of approximately 20 percent, and Churchill believes this rate of growth can be sustained for several years. The 2010 addition of 21 Lunds/Byerly's (lunds andbyerlys.com) stores to the list of retail outlets for Thousand Hills should boost sales considerably. These upscale supermarkets attract affluent shoppers who are interested in high quality local food products.

Churchill identifies human capital as the most significant limit to expansion. As the business grows, he will need to hire more staff. Maintaining service quality for wholesale and institutional customers can be challenging during rapid growth, and hiring and retaining excellent people will be a key to success. Churchill also notes that supply can be a limiting factor. While Thousand Hills pays a significant premium to its producers, raising cattle under the Thousand Hills protocols requires knowledge, attention to detail, and hard work.

Matching sales volume, operational capacity, and business infrastructure is an important management challenge for Churchill. The current Thousand Hills facilities, equipment, and staff are sized to handle some sales growth without major changes. Eventually, however, continued growth will necessitate new facility and infrastructure investments, added staff, and perhaps the development of new business systems and procedures that are better suited for larger sales volume.

Thousand Hills' partnership with Lorentz Meats is critical for continued growth. When asked about significant threats to his business, Churchill immediately noted that an interruption in operations at Lorentz Meats would be highly disruptive, though there are contingency plans in place. Churchill's concern about this potential threat highlights the fact that these two businesses have helped each other grow and that continuation of a strong relationship can benefit both in the future. As noted earlier, Lorentz Meats is already approaching capacity constraints and is considering a significant expansion. This will be costly, and new investment will be justified, in part, by prospects for growth in processing for Thousand Hills.

Finally, Churchill is experimenting with new business ventures that will allow Thousand Hills to use current resources more efficiently and possibly expand to realize new economies of size. One of these ventures, Restoration Raw Pet Food (restorationrawpetfood.com), produces and markets a pet food combining raw meat and meat byproducts with sprouted organic grains and trace minerals. As already noted, Thousand Hills is also providing distribution services for Lorentz Meats' cured pork products and for a local producer's poultry products. This will enable Thousand Hills to use its resources more efficiently and will give customers access to a wider range of products. Churchill is not considering using the Thousand Hills label for these new products.

Neither Churchill nor Mike Lorentz expressed concerns about federal and state food safety regulations. Both view compliance with these as a cost of doing business. Churchill did express concerns about federal commodity policies, stating that feedlots in the industrialized livestock sector are viable because they add value to feed grains that are inexpensive because of federal price support programs. He went on to say: "Until we decide, as a country, that the goal is no longer to add value to corn and soybeans, we will not be able to make a change in the way industrial agriculture works and functions. Thousand Hills Cattle Company is not interested in adding value to corn and soybeans. We are interested in producing high quality, safe meat in an environmentally responsible way." When asked about promotional requirements or other fees to gain access to a store, Churchill indicated that none of the retailers that sell Thousand Hills products had imposed any such requirements. Rather, he noted that reliability is the most important requirement for retailers. They insist on consistency, excellent product quality, proper invoicing, and assurance of correct cold chain management. He went on to say that larger retailers are not always confident that an individual farmer has the expertise to do these things. As an aggregator, he believes that Thousand Hills has the operational capacity to work with accounts that have high demands for operational excellence and consistency.

Mike Lorentz responded to the question about retail access by noting that standards set by major food service distributors—for example, the requirement that ground beef be passed through a metal detector—can be an important barrier for locally produced niche meat products to enter food service distribution channels. Lorentz Meats is positioning itself to comply with these standards, but smaller plants would not be able to afford the investment in equipment and quality control personnel.

Key Lessons

Three general lessons emerge from this case study of an intermediated local food supply chain. First, a strong product brand allows Thousand Hills to develop durable bonds with its customers. Focusing on a consistently high quality fresh product (even at a significantly higher price than other beef products) has allowed Thousand Hills to grow rapidly in a niche market. At the same time, it is noteworthy that the linkages with consumers are not personal, except through in-store demonstrations by producers. Thousand Hills' brand image is built on the company's story rather than on the stories of individual producers. The Thousand Hills web site does not identify a single producer.

Second, as a privately held company, Thousand Hills is able to enforce strict quality standards and maintain supply consistency more easily than can be done by individual producers acting independently or by a producer-owned cooperative (where decisions are made by a group). Churchill has managed growth effectively, strengthening capabilities and capacity along with sales volume. Effective control over distribution has enabled Thousand Hills to manage costs and enter mainstream markets that demand efficient delivery.

Finally, this case illustrates the importance of the processor in a meat supply chain. The story of Thousand Hills is actually the shared story of Thousand Hills and Lorentz Meats. These two companies have grown together and learned from each other, while maintaining their independence. In the process they have contributed significantly to the economy of the communities surrounding Cannon Falls. Looking to the future, they will need to continue their parallel growth despite the fact that they face different barriers to growth. For Thousand Hills, the major challenges will be in maintaining and matching growth in both demand and supply. For Lorentz Meats, the major challenges will be in accessing the significant capital resources needed for plant expansion and in developing a plant design and operating strategy that will allow them to increase efficiency while maintaining their high quality standards.

Cross Case Comparisons

The three case studies conducted for the Twin Cities—Kowalski's, SunShineHarvest Farm, and Thousand Hills—illustrate the richness of alternatives consumers have for purchasing beef products.[10] Taken together, these case studies also show that differences in supply chain structure, size, and performance across supply chain types are more complex and highly nuanced than much of the public discourse on local food systems would suggest.

Supply Chain Structure

The structures of these three supply chains differ considerably. Product changes ownership four times before sale to the consumer in the Kowalski's chain, while it never changes hands in the SunShineHarvest Farm chain and changes hands two or three times in the Thousand Hills chain. Durable trading partner relationships with high levels of trust and information sharing are evident in the Kowalski's chain—where they link Kowalski's, Creekstone, and J&B—and in the relationship between Thousand Hills and Lorentz Meats in the Thousand Hills chain. SunShineHarvest Farm has a strong relationship with the processor, but this relationship does not have the strategic importance seen in the other chains.

As expected, producer prices in the two local chains are not linked to commodity prices. Those in the Kowalski's chain are based on national commodity meat prices, albeit with significant premiums. None of these chains has especially strong linkages to the national industry. On the other hand, SunShineHarvest Farm has benefited from its linkages to the Mill City Farmers Market, Sustainable Farming Association, and the state-sponsored Minnesota Grown Program. Thousand Hills used the Food Alliance Midwest certification program during its initial years of operation, has listings on localharvest.org and eatwild.com, and uses Coop Partners Warehouse. Finally, Thousand Hills' processing partner, Lorentz Meats, has contributed significantly to local foods infrastructure through educational programming, and Thousand Hills has started to offer distribution services to other local food producers.

Supply Chain Size and Growth

These supply chains also differ greatly in size. Thousand Hills markets approximately 32 times more beef animals than SunShineHarvest Farm, yet Thousand Hills' sales represent only about 5 percent of beef sales in a typical Kowalski's store. Thousand Hills beef products are carried in approximately 50 metro area retail stores, but these stores account for less than 25 percent of grocery retail sales in the Twin Cities market. Therefore sales of these two local products account for only a very small percentage of beef sales in the metropolitan area.

Table 18. Allocation of retail revenue in Twin Cities, Minnesota—beef chains, by supply chain and segment

	MAINSTREAM		DIRECT		INTERMEDIATED	
	Kowalski's[a]		SunShineHarvest Farm[b]		Thousand Hills[c]	
Supply chain segment	Revenue ($/lb)	% of Total	Revenue ($/lb)	% of Total	Revenue ($/lb)	% of Total
Producer-finisher	2.39	38.7	3.92	70.8	2.68	37.4
Producer-finisher estimated marketing costs[d]	na	na	0.74	13.3	na	na
Processor[e]	1.73[f]	28.0	0.88	15.9	0.94	13.2
Distributor-Aggregator	na	na	na	na	1.89	26.3
Retailer	2.06	33.3	na	na	1.65	23.1
Total retail value[g]	6.18	100	5.54	100	7.16	100

[a.] We assume a retail value of $3,054 for meat from a whole animal with a live weight of 1,300 lb and a meat yield of 494 lb. Transportation costs from the producer to processor and from the processor to the distributor are borne by the processor. Transportation costs from the distributor to the retailer are borne by the distributor. Distributor-aggregator revenue is combined with revenue accruing to the processor segment to maintain confidentiality.

[b.] We assume a retail value of $2,172 for a whole animal with a live weight of 1,110 lb and a meat yield of 392 lb. This is based on 25 percent of meat being sold in farmers markets and 75 percent of meat being sold through buying clubs or the meat CSA.

[c.] We assume a retail value of $3,040 for meat from a whole animal with a live weight of 1,200 lb and a meat yield of 424 lb. All transportation costs are borne by the aggregator.

[d.] Includes the estimated portion of producer revenue attributed to costs of transport to market, market stall fees, and the opportunity cost of labor devoted to marketing activities. Total per unit revenue for the producer-finisher is 3.92 + 0.74 = 4.66 ($/lb).

[e.] These calculations do not include revenue from processing byproducts.

[f.] The processor value in the mainstream chain also includes distribution costs. For confidentiality reasons, we did not separate these values.

g. Retail values are based on revenue for the meat yield from a whole animal. For mainstream, $3,054 in revenue from 494 pounds of meat; for direct, $2,172 in revenue from 392 pounds of meat; for intermediated, $3,040 in revenue from 424 pounds of meat.

Source: King et al. (2010: 42).

Despite their small size, neither access to processing and distribution services nor fixed costs associated with compliance with regulatory and operating standards appears to be a significant barrier to expansion for the two local chains. Thousand Hills meets product consistency and service quality standards required to be a vendor for several supermarket chains. In contrast, SunShineHarvest Farm would have difficulty meeting product consistency and service quality standards required by many retail outlets, but the Brauchers do not currently have plans to sell in supermarkets.

Finally, the anticipated mode for growth in the two local chains also differs. Major expansions in the supply of direct marketed beef are likely to come through the entry of new, independently managed farm operations, largely due to the lack of economies of size in direct marketing activities. In contrast, Thousand Hills has both plans and the potential for growth through internal expansion, which will likely involve new producers operating under the company's direction.

Supply Chain Performance

The allocation of revenues to producers differs significantly across supply chains, as shown in table 18. SunShineHarvest Farm retains the largest share of consumer expenditures from their direct market enterprises, 70.8 percent, even after netting out processing and estimated marketing costs. In contrast, producers in the intermediated Thousand Hills supply chain retain only 37.4 percent of consumer expenditures, while producers in the Kowalski's supply chain retain 38.7 percent. Producer revenue, net of estimated marketing costs borne by the producer, is higher than the mainstream case for both the direct and intermediated chains and is highest for SunShineHarvest Farm.

Median ground beef prices over 2009 were $3.99, $5.00, and $5.99 per pound for Kowalski's, SunShineHarvest Farm, and Thousand Hills,

Table 19. Food miles and transportation fuel use in Twin Cities, Minnesota—beef supply chains

SUPPLY CHAIN SEGMENT	FOOD MILES	TRUCK MILES
Mainstream Chain Kowalski's[a]		
Cow-Calf to Finisher	250	500
Finisher to Processor	615	1,230
Processor to Distribution	720	1,440
Distribution to Retail	60	120
All Segments	1,645	
Direct Chain SunShineHarvest Farm[b]		
Producer to Processor	20	40
Processor to Distribution	20	40
Distribution to Retail	35	70
All Segments	75	
Intermediated Chain Thousand Hills[c]		
Producer to Processor	250	500
Processor to Distribution	5	10
Distribution to Retail	45	90
All Segments	300	

[a.] All transport in this chain is in semi-trailers that achieve fuel economy of 6 mpg. Live animals are assumed to yield meat with a retail weight of 494 lb. A load of 55 live feeder cattle is transported from the cow-calf operation to the finisher. A load of 40 live cattle is transported from the finisher to the processor. In subsequent segments of the chain, 45,000 lb loads of fresh meat are transported.

[b.] All transport in this chain is in a pickup truck that achieves fuel economy of 16 mpg. Live animals are assumed to yield meat with a retail weight of 392 lb. Three animals are

RETAIL WEIGHT (CWT)	FUEL USE (GAL)	FUEL USE PER CWT SHIPPED
272	83.3	0.31
198	205	1.04
450	240	0.53
450	20	0.04
		1.92
11.8	2.5	0.21
11.8	2.5	0.21
2.5	4.4	1.76
		2.18
115	56	0.49
106	2	0.02
76	14	0.18
		0.69

transported to the processor, and the meat from three animals is transported back from the processor.

c. We assume that a load of 27 cattle born and finished on a farm 250 miles from Cannon Fall MN, transported in a small semi-trailer that achieves fuel efficiency of 9 mpg. Each of these cattle yields meat with a retail weight of 424 lb. All subsequent transportation of meat is in a refrigerated delivery truck that achieves fuel efficiency of 6.5 mpg.

Source: King et al. (2010: 43).

respectively. Median ribeye steak prices for the three supply chains were $13.99, $16.00, and $17.99 per pound, respectively. The SunShineHarvest Farm prices are those posted at the Mill City Farmers Market. Buying club and CSA prices are lower. Products in the farmers market portion of the direct chain and in the intermediated supply chain command a significant premium—ranging from 14 to 75 percent above the ground beef and ribeye products marketed through the mainstream chain. While Kowalski's store brand beef has many of the same qualities as beef marketed through the direct and intermediated chains, it is not grass-fed, nor is it available direct from the production source. These data suggest that consumers are willing to pay more for both of these product attributes. Nearly all the revenue from the direct and intermediated chains remains within the local economy. The only exception is revenue received by cow-calf producers and finishers outside Minnesota and Wisconsin who sell cattle to Thousand Hills. A significant portion of revenue in the mainstream chain accrues to Creekstone and to cow-calf producers and finishers outside the local economy, but a significant portion also accrues to J&B and Kowalski's, both companies that are based in Minnesota.

Products travel fewer miles in the direct and intermediated supply chains (table 19). However, fuel use per 100 pounds of product is highest for SunShineHarvest Farm due to relatively small load sizes. Thousand Hills has by far the lowest fuel use per 100 pounds of product; shorter transport distances offset the inefficiencies of transporting products in loads smaller than the full semi-trailer loads used in the Kowalski's chain. This demonstrates that direct and intermediated supply chains can be efficient when product is aggregated.

Key Lessons

Four general lessons emerge from these case studies of beef supply chains in the Twin Cities.

1. Processing is an essential segment in the supply chain for any meat product. While small-scale processing technology is available for poultry, large animal processing plants require a scale of operation and level of expertise that could not be achieved by either SunShineHarvest Farm or Thousand Hills in their current configurations.

Therefore the availability of processing facilities was an essential precondition for the formation of both of these businesses.

2. Fewer food miles do not imply less transportation fuel use on a per unit basis. This is not a surprising finding. What is noteworthy in our findings is that a local food supply chain can outperform the mainstream chain when it achieves enough volume to be able to take advantage of scale economies in transportation. Results for Thousand Hills suggest that transportation fuel savings can be substantial.

3. Stable trading partner relationships based on transparency, trust, and willingness to forgo short-term gains in order to maintain the benefits of long-term strategic partnerships are seen in all three supply chains. Development of these close relationships is not limited to small firms.

4. The mainstream supply chain is formidable competition for local food supply chains. Kowalski's mainstream supply chain allows them to offer a high quality, differentiated beef product with health and animal welfare attributes valued by consumers at prices consistently below those observed for the direct market and intermediated chains. The two local products in these case studies do have additional attributes for which some consumers are willing to pay a premium, but the local products currently capture only a small part of the overall market for beef in the Twin Cities Metro.

Notes

1. While *local* is defined in this study as food products that originate in Minnesota or Wisconsin, the two local supply chain case studies are Minnesota-based. Therefore this description of the beef sector is confined to Minnesota.

2. Primals are the basic cuts into which an animal carcass is divided for distribution and/or further processing into subprimals or consumer-ready products such as steaks and roasts.

3. Minnesota beef consumption is estimated as Minnesota population multiplied by national per capita beef consumption.

4. All information about Creekstone Farms Premium Beef and its producer, Alan Verdoes, was obtained through secondary sources such as news articles and web sites.

5. "[Processing] establishments have the option to apply for Federal or State inspection. States operate under a cooperative agreement with FSIS. State program must

enforce requirements 'at least equal to' those imposed under the Federal Meat and Poultry Products Inspection Acts. However, product produced under State inspection is limited to intrastate commerce" (USDA-FSIS n.d.).

6. As a reference for comparison with prices for other beef products sold in the Twin Cities, the posted farmers market prices for ground beef and ribeye steak sold by SunShineHarvest Farm are $5/lb and $16/lb, respectively.

7. Name changed to honor confidentiality.

8. Using this estimated cutout and observed retail prices from several locations yielded a similar retail value for a "typical" animal.

9. Kowalski's is the focus for the mainstream beef case, so more detail on their operations is provided there.

10. It is important to note that these three case studies cannot possibly be fully representative of all beef supply chains. Rather, each represents a single instance of a mainstream, direct market, or intermediated supply chain. Mainstream supermarket supply chains may differ considerably across companies, and the structure, size, and performance of mainstream food service supply chains may also be quite different. Similarly, there is considerable variation in operating practices, size, and performance of direct market and intermediated chains.

References

Barnes, G., and P. Langworthy. 2003. "The Per-Mile Costs of Operating Automobiles and Trucks." Technical report #2003–19. St. Paul: Minnesota Department of Transportation.

Bureau of Economic Analysis. N.d. "Regional Economic Accounts." bea.gov/regional/ (accessed 12/21/2011).

Center for Farm Financial Management. N.d. "Livestock Enterprise Analysis: Beef Finish Calves." *FINBIN*. finbin.umn.edu/output/169971.htm (accessed 12/27/2011).

Creekstone Farms. N.d. "Natural Black Angus Beef." creekstonefarms.com/naturalangus beef.html (accessed 12/21/2011).

Davis, Christopher G., and Biing-Hwan Lin. 2005. "Factors Affecting U.S. Beef Consumption." LDP-M-135–02, U.S. Department of Agriculture, Economic Research Service.

El-Osta, H., and M. Ahearn. 1996. "Estimating the Opportunity Cost of Unpaid Farm Labor for U.S. Farm Operators." TB-1848. U.S. Department of Agriculture, Economic Research Service.

Food Institute. 2008. "Food Processing." In *Food Industry Review*. Elmwood Park NJ: American Institute of Food Distribution.

King, Robert P., Michael S. Hand, Gigi DiGiacomo, Kate Clancy, Miguel I. Gómez, Shermain D. Hardesty, Larry Lev, and Edward W. McLaughlin. 2010. *Comparing the Structure, Size, and Performance of Local and Mainstream Food Supply Chains*. ERR-99. Washington DC: U.S. Department of Agriculture, Economic Research Service.

Local Harvest. N.d. LocalHarvest.org (accessed 4/27/2010).

Mill City Farmers Market. N.d. "Sustainability Statement." millcityfarmersmarket.org /about-the-market/sustainability-statement/ (accessed 12/27/2011).

Moran, Sarah. 2009. "Know Your Meat" and "Common Questions about Buying Direct." *Star Tribune,* July 30. startribune.com/lifestyle/taste/51988697.html?elr=KArksUUUoDEy 3LGDiO7aiU and startribune.com/lifestyle/taste/51988757.html?elr=KArksUUUoDEy 3LGDiO7aiU (accessed 12/27/2011).

National Cattlemen's Beef Association. N.d. "Beef Cutout Calculator." beefresearch.org /CMDocs/BeefResearch/Beef%20cutout%20calculator.pdf (accessed 12/21/2011).

Roti, Lura. 2008. "Creekstone Farms Premium Beef Offers More Marketing Options." *Tri-State Neighbor,* December 11. tristateneighbor.com/articles/2008/12/11/livestock _guide/lsg02.txt (accessed 4/27/2010).

Sustainable Farming Association of Minnesota. N.d. "About Us." sfa-mn.org/about/ (accessed 12/27/2011).

Thousand Hills Cattle Company. N.d. "100% Grass-Fed Beef Program Protocol." thousand hillscattleco.com/files/ProducerProtocol2008.pdf (accessed 12/27/2011).

USDA-ERS. N.d. "State Fact Sheet: Minnesota." ers.usda.gov/StateFacts/MN.htm#PIE (accessed 4/27/2010).

USDA-FSIS. N.d. "Regulations and Policies: State Inspection Programs." fsis.usda.gov /regulations_&_policies/state_inspection_programs/index.asp (accessed 12/27/2011).

USDA-NASS. 2008. "Livestock Slaughter 2007 Summary." usda.mannlib.cornell.edu /usda/nass/LiveSlauSu//2000s/2008/LiveSlauSu-03-07-2008_revision.pdf (accessed 12/21/2011).

———. 2009. *Census of Agriculture, 2007.* Washington DC: U.S. Department of Agriculture, National Agricultural Statistics Service.

7 Fluid Milk Case Studies in the Washington DC Area

Michael S. Hand and Kate Clancy

Introduction

This set of case studies describes the distribution of fluid milk through three different supply chains in the Washington DC metropolitan area:

private-label milk that is processed by a producer cooperative and sold in mainstream supermarkets,
direct farm-to-consumer home delivery, and
locally produced milk that is sold at a small chain of organic markets.

The Location: Washington DC and Surrounding Area

Washington DC is the hub of a major metropolitan area in the mid-Atlantic region. We consider supply chains that primarily serve customers in the core of the Washington DC area, within either the District of Columbia proper or its neighboring counties. For these case studies a food product is considered local if it is produced within the counties comprising the Washington–Baltimore–Northern Virginia combined statistical area defined by the U.S. Census Bureau, plus the counties that share a border with the combined statistical area. The estimated total population of the area defined here was approximately 10.5 million in 2009 (U.S. Census Bureau 2009).

This area, subsequently referred to as the DC area, contains the District of Columbia and 56 counties across five states (Maryland, Delaware, Pennsylvania, Virginia, and West Virginia; see figure 21). By including the

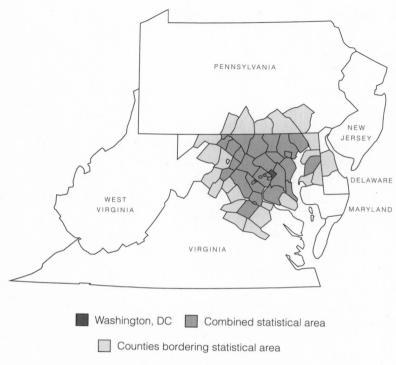

Washington, DC █ Combined statistical area

☐ Counties bordering statistical area

Fig. 21. Geographic definition of the Washington DC local area.

counties adjacent to the combined statistical area, the local area captures geographic areas that are important for the three supply chains studied here. In particular, counties in southeastern Pennsylvania traditionally have a relatively high density of dairy farms and supply milk to the DC area through a variety of distribution channels.

The DC area has a vibrant local foods community that includes a variety of outlets for locally produced foods. It has a total of 177 farmers markets (USDA-AMS n.d.) and 4,009 farms that sell directly to consumers with sales of $49.8 million in 2007 (USDA-NASS 2009).[1] Several of the states represented in the DC area maintain state product promotion programs (e.g., Maryland's Best, West Virginia Grown, and Virginia Grown), and the DC Food Finder (dcfoodfinder.org) includes local food outlets in its interactive map of food resources in the District of Columbia.

The Product: Fluid Milk

Although dairy farms and retailers sell a wide variety of dairy beverages and products, we focus on supply chains for fluid cow's milk (referred to as simply fluid milk or milk in these studies). Fluid milk products share many common characteristics, such as the availability of particular fat contents (whole, 2 percent, skim) and sizes (primarily half gallons and gallons). But a wide variety of other characteristics differentiate brands, producers, and retailers. Milk products in these supply chains include conventional and organic milk, milk produced without the hormone rBST, varying degrees of grass-based and grain-fed production, and different packaging materials (plastic, cardboard, and glass).

Per capita consumption of fluid milk has steadily declined over the last several decades relative to other beverages, yet milk is still a staple food in the typical American household. Total plain fluid milk (whole and reduced-fat milks, not including flavored milk or buttermilk) available for consumption in 2008 was about 163 pounds per capita in the United States, or about 19 gallons per person (USDA-ERS n.d.). Fluid milk for consumption represents about 26 percent of all milk produced in the United States.

Milk Supply in the DC Area

Milk purchased in the DC area comes from a variety of sources, both nearby and across the country. The area also produces milk that is distributed and consumed across a large area. The DC area contains a dairy herd of more than 250,000 cows (USDA-NASS 2009: table 11) that produces about 4.7 billion pounds of milk per year (USDA-NASS 2010).[2] Production in the area is concentrated in Lancaster and Franklin Counties in Pennsylvania. These two counties contain more than half of the area's entire dairy herd.

Most of the DC area is contained within the Northeast Area (Order Number 1) of the Federal Milk Order (FMO). The FMO sets minimum prices for the different classes of milk processed through plants and facilities regulated by the FMO. There are 18 plants in the DC area that are part of the Northeast FMO. These plants produce a variety of dairy products, including fluid milk, creamery products (e.g., ice cream and butter), and powdered milk.

This study of fluid milk took place during one of the worst dairy crises in U.S. history. Reduced world and U.S. demand as a result of the recession caused the prices paid to farmers to decrease 50 percent from the year before, while greatly increased feed prices between 2007 and 2009 simultaneously increased operating costs for dairy farmers. Although consumers did not see a large drop in retail prices, farmers received prices well below the cost of production for much of the year 2009. Prices paid to farmers began to rebound at the end of 2009 and beginning of 2010.

Sales of fluid milk in the DC area are dominated by a few large retail supermarket chains. Three chains control more than half of the retail food store market in the area. All are owned by large grocery companies with corporate headquarters outside the region. Milk is predominantly sold in these stores through private-label brands, although each chain carries a varied line of milk products and brands, as do other stores in the area.

The average retail price for a gallon of whole milk in the DC area was $3.30 in October 2009 (when the case studies were conducted) compared to the national average of $2.78 per gallon (California Department of Food and Agriculture n.d.). Based on per capita consumption of plain fluid milk in the United States, it is estimated that consumers in the DC area consume about 1.7 billion pounds of milk each year (or about 195 million gallons).

Mainstream Case: Maryland and Virginia Milk Producers Cooperative

The Maryland and Virginia Milk Producers Cooperative Association (mdvamilk.com, called Maryland and Virginia Co-op or simply the Co-op) produces and supplies private-label milk to supermarket retailers in the DC area. Milk sold under a private-label supermarket brand is common in the northeastern United States, including the Washington DC area. In a study from 2005, between 71 percent and 85 percent of milk sold in supermarkets in the Northeast is marketed as a private-label brand (Bonanno and Lopez 2005).

Supply Chain Structure, Size, and Performance

This study describes the supply chain for white fluid milk produced, processed, and distributed by the Maryland and Virginia Co-op (figure 22). Based in the DC area, the Co-op represents about 1,500 farms in

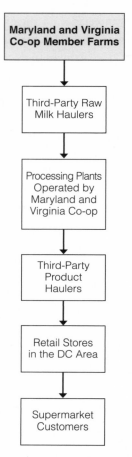

Fig. 22. Mainstream supply chain, Maryland and Virginia Co-op.

11 mid-Atlantic and midwestern states, with about 1,000 farms located in Maryland and Pennsylvania. In total, Maryland and Virginia Co-op processes and distributes about 7 million gallons of milk per month in the DC area (Dudlicek 2009). The majority of milk produced by Maryland and Virginia member farms is produced without the hormone rBST.

The Co-op operates several plants in the mid-Atlantic region. Two of these plants process milk in the DC area for sale in area supermarkets. In addition to processing and packaging milk for private-label customers, Maryland and Virginia Co-op also operates a creamery that produces butter and ice cream and has balancing operations to produce milk powder and condensed milk (Dudlicek 2009). Operating multiple plants and offering several product lines provides the Co-op with flexibility in managing production for a large volume of a highly perishable product.

Milk is typically picked up from member farms by third party haulers or by Co-op-owned trucks. Semi-trailer milk tankers with a gross vehicle weight rating greater than 33,000 pounds are used to assemble milk and deliver to processing plants operated by Maryland and Virginia Co-op. Routes are planned and scheduled to maximize the size of tanker loads to the processing plant and to minimize distance traveled. Information systems at the plant level allow milk at any stage of processing to be traced to a particular tanker load. Because the Co-op's member farms are concentrated in states that contain portions of the DC area, and two of the plants are within the DC area, it is likely that most private-label milk produced by the Co-op is sourced from within or near the DC area.[3]

Co-op members receive prices that are based on the Federal Milk Order for the Northeast Area. From September through November of 2009 the average price of raw milk was about $0.64 per half gallon.[4] Average production costs for dairy farms in the region ranged from $0.63 per half gallon to $0.66 per half gallon, indicating that during the study period the average farm in the mainstream supply chain for milk had recently received prices just covering production costs.[5]

The median retail price of private-label milk at selected supermarkets in the DC area was about $1.99 per half gallon.[6] Because it is associated with a supermarket brand, private-label milk generally conveys little information about where and by whom it was produced. This holds true in the DC area, where information on labels and at the point of sale for the major private-label brands displays only the location of the processing plant or distribution center.

Community and Economic Linkages

The mainstream private-label milk supply chain supports significant economic activity at the local and regional level. Some portion of all the activities in the supply chain is carried out within the DC area, including farm operations, processing and packaging, distribution, and retailing. Given the volume of milk distributed through the supply chain, the total economic impacts are likely to be significant and accrue mostly within the DC area and immediate surrounding areas.

Actors in the supply chain do not appear to be engaged in community events or civic organizations on a wide scale. However, it is common for supermarket retailers in the area to participate in community-wide charitable activities.

Prospects for Expansion

As a well-established mainstream supply chain, expansion is expected to be governed by broad market forces that affect milk demand and the supermarket industry in the DC area. Taken as a whole, the private-label milk market in the area is likely to expand only if there is a general increase in demand for milk. This could occur through an increase in population or a sustained increase in per capita milk consumption. Recent trends, however, suggest that milk market share is being eroded by other beverages.

The Maryland and Virginia Co-op, operating in a competitive marketplace, may face different prospects for expansion compared with the market as a whole. The Co-op could expand if the customers it supplies capture additional market share relative to retailers that it does not supply. The Co-op has positioned itself to operate as a mainstream provider of milk produced without the use of the hormone rBST, and this may allow it to differentiate its products to some degree. But milk supply in the region in which Maryland and Virginia Co-op operates is also competitive, and it is possible that other milk cooperatives and processors could expand at the expense of the Co-op's business. For example, some independent processors may be able to pay slightly higher premiums for raw milk, resulting in market power in the selection of supplier farms. During a period of sustained lower prices or reduced milk demand, these forces may constrain Maryland and Virginia Co-op's ability to expand.

Policy and Regulatory Issues

Two major policy issues affect this milk supply chain—FMO prices and food safety regulations— although neither appears to be a barrier to growth or to impede operations in the supply chain. In fact, the supply chain is designed to operate successfully in a highly regulated and structured market.

Although some individual farms are negatively affected when milk prices decrease substantially (as happened during 2009), the FMO likely has an overall impact of providing stability to milk production and processing in the area. Farms that could not gain substantial premiums for their milk outside a cooperative or through differentiation (e.g., for organic or other quality-differentiated characteristics) may be a particularly good fit for cooperative membership and the FMO price regime.

The regulation of milk production and processing is largely carried out by individual states. For this supply chain, plants in the DC area operated by the Maryland and Virginia Co-op are subject to the Maryland Division of Milk Control microbiological standards and inspections. No known problems with product quality or safety have been observed for these plants.

Key Lessons

Two key lessons emerge from the Maryland and Virginia Co-op supply chain. First, the supply chain has developed in order to efficiently provide supermarket retailers with a stable supply of a basic grocery food staple. Much of this stability is facilitated by the Co-op. As a vertically integrated cooperative, Maryland and Virginia Co-op controls the milk and the processing operations until the product leaves the plant for retail stores. This arrangement is particularly well suited to serving retailers that demand consistent quality, service, and price.

Second, although private-label milk brands typically do not reveal significant information about their origins (aside from the location of the processing plant or distribution center), milk handled by the Co-op may be considered a local food product based on the geographic location of its suppliers. Many of the supplier farms are located within the local production area for DC defined in these case studies, or in nearby counties that are traditionally associated with the region's dairy production. Private-label milk is often marketed as a loss leader for milk within supermarket chains, so it is unlikely that retailers will attempt (or be able) to capitalize on the locality of the product's origin by charging a price premium.

Direct Market Supply Chain: South Mountain Creamery

This study describes the supply chain for milk sold directly from a dairy farm to consumers in the DC area. South Mountain Creamery (southmountaincreamery.com) is a milk producer and processor located near Middletown in Frederick County, Maryland, that bottles and sells its own milk through direct home delivery. Dairy operations began in 1981 under the ownership of Randy and Karen Sowers, and milk processing and home delivery began in 2001.

Prior to 2001 South Mountain was a traditional dairy operation selling milk through a cooperative dairy association. The Sowers decided to operate the creamery as a way to expand their business and capture a larger share of the retail value of their milk. The original vision of the business was to deliver directly to 1,000 homes in the Frederick County area. Today South Mountain delivers to about 4,000 accounts in the greater Washington DC metropolitan area. The vast majority of these customers are outside Frederick County but within a 70-mile radius.

South Mountain's primary line of business is the milk they produce, process, bottle, and sell to home-delivery customers. In addition to milk products, they sell creamery products made at the South Mountain processing plant (e.g., butter, yogurt, and ice cream), meats, cheeses, other beverages, jams and jellies, and other specialty products sourced from a variety of nearby farms and food distributors. South Mountain also operates a small retail store and ice cream counter attached to their processing plant and sells at four area farmers markets.

Supply Chain Structure, Size, and Performance

As a direct home-delivery business, South Mountain is the only major actor in this supply chain (see figure 23). All aspects of production, processing, and retail delivery are controlled and coordinated by South Mountain. There are a handful of small wholesale accounts that resell South Mountain products, but this accounts for a minor portion of South Mountain's business. This description focuses on how South Mountain produces and distributes milk through home-delivery operations.

Production

South Mountain maintains a milking herd of about 220 cows, mostly Holsteins and a few Brown Swiss mixed breed cows. The farm operates on a total of about 1,400 acres, with 80 acres of pasture that is used as a feeding option for the herd. Cows are not confined to pens or barns and

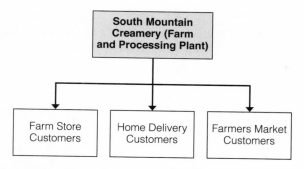

Fig. 23. Direct market supply chain, South Mountain Creamery.

can choose whether to feed from available pasture or the grain ration provided. Cows are also provided with a grain feed ration of 85–90 percent wet silage and 10–15 percent grain and minerals. Aside from the mineral component, all feed is grown on the farm.

Total feed requirements for the farm are about 28,000 pounds per day. When pasture availability is at its peak, the necessary feed ration can drop by between 7,000 pounds and 10,000 pounds per day. Milk production per cow is also lowest during months of peak pasture availability. South Mountain does not use the hormone rBST with its herd. Cows are milked twice each day, with milk being stored in an on-farm tank.

Processing and Packaging

Raw milk is pumped daily via an underground pipeline from the milking parlor storage tanks to a storage tank in the adjacent creamery. First, milk for separation is pumped to the creamery to process creamery products and reduced fat milks. Then the remaining amount of milk needed for whole milk processing is pumped to the creamery. Milk is processed four days per week using high-temperature, short-time (HTST) pasteurization at 168 degrees Fahrenheit. South Mountain's equipment can pasteurize about 600 gallons per hour at a cost of about $1.00 per half gallon.

Pasteurized milk is kept in a holding tank prior to bottling. South Mountain bottles half-gallon and quart containers of skim, 2 percent, homogenized whole milk, and non-homogenized (creamline) whole milk. All fluid milk products are bottled in returnable glass bottles that are washed at the creamery. In total South Mountain processes about

23,000 gallons of milk per month. With their current equipment and facility, South Mountain could increase milk processing by as much as 50 percent without further investments. The plant employs about 22 production workers and managers with an estimated annual wage bill of about $636,000.

Marketing and Distribution

These functions are in three categories.

HOME DELIVERY: About 85 percent of South Mountain's sales are through their home-delivery operation. Customers place orders through the South Mountain web site by logging into to their customer account. Most customers have a standing weekly or semi-weekly order that they receive without having to place a new order each week. Through the web site customers can make adjustments to standing orders, if necessary, at least two days prior to delivery.

South Mountain has about 4,000 home-delivery accounts, of which about 3,200 receive orders in any given week. Deliveries are made five days per week (Monday through Friday) beginning as early as 3:00 a.m. and finishing by about 5:00 p.m. Delivery vehicles are Chevrolet P-30 Step Vans that were purchased used. Each delivery is left in a cooler or box on a porch or front step or in some cases left in garage refrigerators. Deliveries are generally not made to buildings or apartment addresses where stairs or elevators would require additional delivery time.

South Mountain does not have a minimum order size. The size of orders varies from as little as one quart of milk (at $2.25) to $250 per week, although each delivery is charged a flat fee of $3.75. Most South Mountain milk is sold in half-gallon containers at a price of $3.25 per half gallon.[7] Each order generates an invoice that is adjusted based on actual deliveries and bottle returns ($1.50 per bottle). In the past, customers have paid by check, with drivers handling payments. South Mountain has begun switching to credit card billing to avoid having drivers handle checks and deal with collections. However, credit card billing is not required, and many customers still pay by check.

South Mountain employs 13 full-time delivery drivers who operate 52 delivery routes per week, traveling 9,100 vehicle miles per week on average. Marketing costs for the home-delivery operation, including transportation fuel, vehicle maintenance and depreciation, and driver wages, total about $1.03 per half gallon.[8]

FARMERS MARKETS, FARM STORE, AND WHOLESALE ACCOUNTS: In addition to home-delivery sales of milk and other products, South Mountain sells milk at four farmers markets in Maryland and Virginia, operates a small farm store, and sells to a handful of wholesale accounts. In total, these lines of business account for about 15 percent of South Mountain's sales.

Although they represent a relatively small portion of sales, the additional outlets are an important part of South Mountain's business. The farmers markets give South Mountain a sales outlet during the summer months when demand for milk through home delivery slackens. The farm store is also a draw for families and groups who visit the farm. South Mountain allows visitors to feed calves and view the milking parlor operations, and they host bi-annual farm festivals where they give tours of the processing plant. These marketing outlets may also serve as a venue for contacting new customers.

TRADITIONAL MILK MARKETING: Prior to operating the creamery and processing plant, the farm at South Mountain sold milk through a traditional cooperative milk marketing association, the Maryland and Virginia Milk Producers Cooperative that is the focus for the mainstream case study. South Mountain continued to sell excess milk through the Maryland and Virginia Co-op after creamery operations began, selling up to 1,100 gallons per week. South Mountain would be paid for these sales based on the Federal Milk Order (FMO) for the Northeast Area blend price, between $12 and $13 per hundredweight at the time of the study in 2009. However, increased production resulted in South Mountain being reclassified as a pool producer and receiving substantially less for their milk. The lower prices led South Mountain to cease milk marketing through the cooperative, and South Mountain now uses excess milk to produce its own additional byproducts, such as butter.

Food Miles and Transportation Fuel Use

In this supply chain, milk travels directly from the farm to the home with no intermediary transportation. Thus, the "food miles" of South Mountain milk are relatively low. Most delivery customers live within a

70-mile radius of the farm, with many customers within Frederick County, Montgomery County, and the northwest quadrant of Washington DC. The typical distance milk travels in the supply chain is estimated to be about 48 miles.[9]

South Mountain delivers five days per week using 12 delivery routes on one day and 10 routes on the remaining four days (52 routes total). Total roundtrip mileage for each delivery route averages about 175 miles, including roundtrip travel to the delivery area and distance between deliveries. Delivery trucks typically get about 10 miles per gallon, resulting in total fuel use (gasoline) for deliveries of about 17.5 gallons per round trip. Non-dairy products account for a portion of this fuel use, although about 90 percent of the space and weight on a typical delivery route is accounted for by milk. Apportioning 90 percent of fuel use to milk deliveries, South Mountain uses an estimated 1.9 gallons of gasoline per hundredweight of milk sold.

Community and Economic Linkages

South Mountain's home-delivery business yields direct economic impacts in the form of wages and salaries paid to plant employees, delivery drivers, office employees, and managers. It is likely that almost all employment supported by South Mountain can be directly attributed to the switch from cooperative marketing to on-site processing and home-delivery operations. Some of the additional employment may also be supported by South Mountain's distribution of non-dairy products that generate larger orders and more delivery routes.

Although the activity is not quantified here, South Mountain is also responsible for indirect economic impacts through its wholesale purchasing and distribution of non-dairy products. Many of these products are purchased from distributors and sourced from across the United States and other countries. However, some products are sourced from farms and operations in the Frederick County area. To the extent that South Mountain provides an outlet for these products that would not be otherwise available, some employment and income are indirectly supported by South Mountain's operations.

South Mountain's farm and farm store may support community and civic engagement through farm visits. South Mountain hosts two farm open house festivals, one in spring and one in fall, which offer visitors a chance to tour the milk processing plant. South Mountain also occasionally hosts school groups on farm visits. Aside from the bi-annual festivals, farm visits tend to be informal and periodic.

The festivals and farm visits likely serve a primary purpose of communicating South Mountain's values and fostering trust with its customers. No events are currently conducted in areas where customers live, nor are there any that involve other nearby farms that supply products for South Mountain's delivery business. South Mountain does not appear to participate formally in any civic organizations in the Frederick County area or in the greater Washington DC area.

Farmer Profitability

Fluid milk sales are the core of South Mountain's home-delivery business, but assessing performance based on farmer profitability requires consideration of the other lines of business. Selling non-dairy products that are not produced on the farm is likely a key to success and profitability. Interviews and separate estimates suggest that profit margins are narrow for fluid milk but larger for the products sourced from other farms and distributors. In addition to the larger margins, the full line of product offerings appears to encourage customers to make more regular orders and appeals to a wider range of customers.

An estimate of marketing margins under current operations compared to sales of an equivalent volume of milk through cooperative marketing suggests that current operations may be more profitable for South Mountain. Revenues from cooperative marketing at $13 per hundredweight would result in gross revenues of $6,000 per week at current production levels.[10] Using current prices from the South Mountain web site for half gallons ($3.25) and quarts ($2.25) of milk, gross revenues from milk are estimated at $38,208 per week with direct home delivery (including delivery costs). Processing costs and the milk portion of delivery and other costs are estimated to be about $29,600 per week.[11] If current overhead and farm input costs are similar to what they were before the home-delivery operation began, the gross milk marketing margin appears to be about $2,600 per week greater using direct delivery than traditional cooperative marketing. This gross margin would be reduced if there are costs associated with debt service for new equipment and facilities acquired to process milk and deliver to customers.

South Mountain's current operation involves more than just fluid milk sales. Sales of non-milk products from other farms and distributors make up as much as 63 percent of total revenues, and these products are important for attracting business and boosting retail margins. Non-milk product sales are estimated to contribute up to 80 percent of South Mountain's weekly gross marketing margin.

Prospects for Expansion

South Mountain would like to expand from 4,000 to about 5,000 customers within their current delivery area. Equipment installed recently could accommodate an increase in production of up to 50 percent if necessary. South Mountain currently has some surplus milk it is processing as butter and other creamery products, and they could accommodate some growth in fluid milk demand using this surplus. However, other physical constraints (e.g., land base for feed crops and pasture to support additional production) may limit growth in the near term without significant investments.

Expansion of the business would likely require additional investments in refrigerated storage space and product handling equipment (e.g., conveyors to transfer filled milk crates to the cooler). Further, product handling and inventory controls would need to be improved to handle more orders and additional product offerings. Currently delivery drivers assemble and load orders for their delivery routes, which would become less efficient with the addition of more routes. Inventory control can also be difficult as South Mountain distributes up to 300 product codes from its dairy products and those received from other farms and distributors.

Policy and Regulatory Issues

As with most supply chains for milk, pricing and food safety and inspection regulations are the primary policy issues of interest. The FMO rules have classified South Mountain as a pool producer rather than a producer-handler. This has resulted in a lower price for extra raw milk that it would like to sell through the Maryland and Virginia Co-op and has effectively eliminated this potential sales outlet. Given that these rules are based on production volume, it is unlikely that South Mountain will begin selling raw milk through the cooperative unless the exemption rules are changed.

Milk safety and inspection regulations have generally not been a problem for South Mountain, although there was a steep learning curve when they began to process and market their own milk. The most difficult aspect of this process was learning about all of the necessary permits and fees that are required. The Maryland Division of Milk Control has been helpful in coordinating the permitting and regulatory process, and South Mountain is able to go to the Division of Milk Control to troubleshoot production and regulatory problems. South Mountain is one of the first operations of its type in Maryland, so it is likely that many of their regulatory issues are not commonly faced in the state.

Key Lessons

South Mountain has built a unique local food business that supplies milk, dairy, and other products to more than 4,000 households in the DC area. The business began as an opportunity to capture a greater share of the value of the dairy farm's product, and sales directly to consumers have completely supplanted traditional milk marketing operations for the farm.

Several key lessons about direct-to-consumer marketing are apparent. First, the supply chain in this case is short, both in terms of geographic distance (70 miles or less) and the number of supply chain segments, but the business nonetheless requires complex logistical operations and product handling. In addition to fluid milk and dairy products, South Mountain sells up to 300 products from distributors and other nearby farms. Inventory control and delivery logistics become increasingly complicated with the addition of products and customers, and growth in the delivery business has required significant skill to manage the added complexity.

A second lesson is that although milk is the focal product for South Mountain, the non-dairy products sold through the home-delivery business are key to the company's success and profitability. Margins on the non-dairy products tend to be larger than for milk, so these products are an important revenue stream. Also, carrying a line of diverse products appeals to a wider variety of households and encourages home-delivery customers to place regular orders. Expansion in the future will rely on

South Mountain's ability to carry products that customers demand and that fit within South Mountain's market niche (i.e., food products delivered directly to customers).

The shift in South Mountain's business from cooperative milk marketing to home delivery has involved significant entrepreneurial activity, and the owners and managers have built entrepreneurial skills as they have expanded the business. South Mountain has essentially started several lines of business from scratch, including milk processing and bottling, creamery product manufacturing, logistics and distribution, and food retailing. None of these lines of business is itself particularly unusual (e.g., milk home delivery is the "old way" of milk distribution), but as a combination they have been designed to fit a particular market niche.

Finally, it appears that South Mountain is operating as the hub of a unique food distribution network. South Mountain sells products from several nearby farms and an independent distributor through its home-delivery service. Although milk products are the backbone of the business, they are increasingly relying on their expertise in distribution and delivery logistics to expand. South Mountain's investments in delivery infrastructure (i.e., delivery trucks and refrigerated storage) allow them to serve essentially as a multi-product aggregator and retail outlet for area farms. This ability may be a key to expanding the market for farm products produced in the Frederick County area and sold directly to DC area households.

Intermediated Supply Chain: Shankstead EcoFarm– Trickling Springs Creamery–MOM's Organic Market

This study describes the supply chain for milk distributed by Trickling Springs Creamery (tricklingspringscreamery.com) and sold at MOM's Organic Market (momsorganicmarket.com) retail stores in the DC area. Trickling Springs operates a milk processing plant and store in Chambersburg, Pennsylvania, which is just under 100 miles from downtown Washington DC. Trickling Springs sells organic and natural milk and other dairy products to wholesale customers under its own label and under private-label agreements. MOM's carries Trickling Springs milk under its own private label and in Trickling Springs-labeled returnable glass bottles at its five stores. MOM's is Trickling Springs' largest private-label customer.

Milk processed at Trickling Springs is sourced almost exclusively from Shankstead EcoFarm, an organic, grass-based dairy farm about nine miles from the Trickling Springs plant. Shankstead does not sell to any other processors. Shankstead also bottles on farm and sells some of its milk as raw (unpasteurized) milk under its own brand, The Family Cow. As a whole, Trickling Springs milk, and thus Shankstead milk, is sold in more than 150 locations in the Washington DC metropolitan area and in Pennsylvania.

Trickling Springs milk is likely considered to be a local product by MOM's customers and at other retail stores, and it meets the definition of local products used in the Washington DC case studies. Trickling Springs and Shankstead work together to inform wholesale customers of the source of the milk and how it is produced, which is a valuable selling point. Also, MOM's prefers to have private-label milk that is sourced from a nearby processor and farm.

Supply Chain Structure, Size, and Performance

Trickling Springs is the hub of this supply chain. They pick up milk from Shankstead, deliver it to the Trickling Springs–owned plant for processing, and deliver finished products to its wholesale accounts (see figure 24). This section describes how milk moves through the supply chain from the farm into the hands of MOM's customers.

Production

Milk processed by Trickling Springs comes from Shankstead, a 250-head dairy farm run by Edwin Shank and his family. Shank's operation is meant to mimic natural processes and animal habits. The cows' diet is primarily grass-based, and the herd is rotated through 120 acres of pasture that is divided into 28 paddocks. The herd also receives some grain-based feed as a supplement. During peak pasture production, grain feed is limited to about one to two pounds of grain per cow per day, although the supplement can be as high as 10 pounds per cow per day in the winter.

Most of the feed grain is whole organic corn purchased from a farm in Virginia. The corn is roasted and ground at Shankstead. Some additional

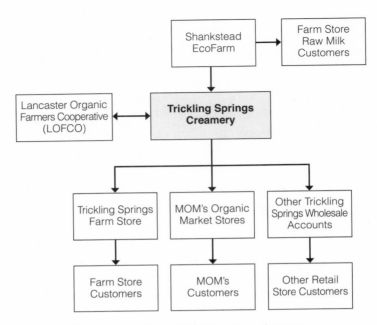

Fig. 24. Intermediated supply chain, Trickling Springs Creamery.

organic corn is purchased for feed as silage. In addition to purchased organic corn, Shankstead grows grass and legumes on 200 acres for winter forage, typically producing about 12 tons of high-moisture feed per acre per year.

In addition to the mostly Jersey dairy herd, the Shank family raises about 600 layer hens for eggs and about 500 broiler chickens. The chicken flock is meant to complement the cows' pasture rotation, with movable pens that are rotated through the pasture a few days after the cows have grazed. This allows the chickens to engage in "pasture sanitation," where the birds feed on insects and help spread manure more evenly in the paddock to increase pasture growth.

Milking is done once a day, with milk stored in a tank on the farm. Cows produce about 40 pounds (4.65 gallons) of milk per day on average. In total, Shankstead produces about 3,000 hundredweight of milk per month, or about 35,000 gallons. Shankstead employs three full-time workers with a total annual wage bill of between $115,000 and $120,000, although their time is not devoted solely to milk production. Family labor also helps to run the various farm operations and the farm store.

Although Shankstead is currently the only raw milk source for Trickling Springs, Trickling Springs is able to purchase milk from the Lancaster Organic Farmers Cooperative (LOFCO) to meet short-term needs. Trickling Springs can also sell extra milk to LOFCO if they cannot use all the milk they pick up from Shankstead. Currently Trickling Springs has a surplus of milk from Shankstead, which is being sent to LOFCO.

Raw milk prices received by Shankstead are determined by contractual agreement with Trickling Springs, which is revised every six months. As of October 2009 Shankstead received about $30 per hundredweight, or about $1.29 per half gallon. Prices are based on average prices paid for organic milk by other major organic milk brands in the region. In addition to a base price, Shankstead may receive additional premiums based on quality.

Processing and Packaging

Trickling Springs picks up raw milk from Shankstead four days per week in a Trickling Springs–owned milk tanker. Milk is processed four days per week, pasteurized using the high-temperature, short-time method, and bottled in either glass or plastic bottles. As of August 2009 Trickling Springs was processing about 2,300 hundredweight of milk per month but occasionally up to 2,600 hundredweight per month. Trickling Springs has a small cooler for storage of finished products. The storage space is limited, however, so inventory is turned around for delivery relatively quickly.

Trickling Springs employs a total of about 35 people, of whom approximately 30 are involved in milk production, delivery, or management. Wages for these workers range from $10 per hour to $16 per hour, for an estimated annual wage bill of about $780,000.[12] The remaining employees work in the Trickling Springs retail store.

Trickling Springs purchases both plastic and glass bottles from third party vendors. Although the glass bottles are sold at retail with a bottle deposit, only about 25 percent of their glass bottles are returned. Trickling Springs uses a caustic-wash machine to wash returned bottles.

Marketing and Distribution

The focus of this supply chain study is Trickling Springs milk that is sold in the five MOM's stores in the DC area. MOM's is an important wholesale customer for Trickling Springs. About 17 percent of the milk Trickling Springs processes is sold as fluid milk in MOM's stores, or about 386 hundredweight per month.

Each MOM's store places individual orders with Trickling Springs and receives deliveries twice weekly. Orders are placed either by phone or by fax. MOM's orders private-label milk in gallon and half-gallon plastic bottles and Trickling Springs–labeled milk in half-gallon glass bottles. Wholesale prices are the same for all Trickling Springs customers, including MOM's, at $3.11 per half gallon. Trickling Springs delivers to MOM's (and other stores) using Trickling Springs–owned Freightliner delivery trucks. In total Trickling Springs operates up to five delivery trucks per day, with the average roundtrip delivery route totaling about 250 miles.

The five MOM's stores are between 50 and 100 miles from the Trickling Springs plant. Three of Trickling Springs' delivery routes serve the MOM's stores on Tuesday and Friday each week, along with several other wholesale accounts in the DC area. MOM's milk products on the three Tuesday routes account for about 33 percent, 25 percent, and 11 percent of the milk on the respective delivery trucks. On the three Friday routes, MOM's milk products account for about 50 percent, 55 percent, and 40 percent of the milk on the respective delivery trucks. The average load share for MOM's milk products across the six delivery routes is 36 percent.

The private-label milk from Trickling Springs is an important part of MOM's milk business because it is the only milk product they sell in gallon containers. By volume, milk in gallon containers accounts for 38 percent of all milk sold at MOM's stores, with an additional 12 percent accounted for by private-label milk in half gallons and another 12 percent from Trickling Springs–labeled milk in glass bottles. In total, milk processed and packaged by Trickling Springs makes up about 57 percent of the volume of milk sold at MOM's stores. As of October 2009 MOM's also sold the milk brands Natural by Nature and Organic Valley.

Private-label milk may be considered a "loss leader" for MOM's. The price of a half gallon of MOM's private-label milk in October 2009 was $3.29. Profit margins for MOM's private-label milk are narrow (and possibly negative), and lower prices tend to support demand for private-label milk as compared to the national brands. Glass-bottled Trickling Springs milk, on the other hand, fills a niche market for MOM's stores. Although this product accounts for only about 13 percent of sales, many buyers of glass-bottled milk shop at MOM's specifically for this product.

Food Miles and Transportation Fuel Use

Product in this supply chain travels from the farm to the processing plant (about nine miles) and from the plant to the five MOM's retail stores. The average distance from Trickling Springs to the MOM's stores is about 85 miles, resulting in food miles in this supply chain of 94 miles. All of the vehicles in this supply chain are heavy-duty trucks that get about 5 miles per gallon of diesel fuel.

Roundtrip travel to deliver raw milk from Shankstead to Trickling Springs (18 miles) requires about 3.6 gallons of diesel fuel, or 0.02 gal/cwt of milk shipped. The average roundtrip distance of delivery routes from Trickling Springs that serve the five MOM's stores is 250 miles. Each of these delivery routes also serves other wholesale accounts. The average share of the loads that is attributable to MOM's deliveries is about 36 percent, which is used to apportion the share of fuel use to the MOM's deliveries. Fuel use for deliveries from Trickling Springs to the MOM's stores is about 31 gallons of diesel fuel, or 0.76 gal/cwt of milk shipped. In total the supply chain uses 0.78 gallons of diesel fuel for every hundredweight of milk shipped to MOM's stores.[13]

Community and Economic Linkages

Economic linkages in the supply chain are illustrated by considering the change in employment and wages that would occur under circumstances of supply chain failure. Three hypothetical scenarios are considered in turn—discontinuation of the Shankstead milk business, closure of Trickling Springs' milk processing, and MOM's discontinuation of orders from Trickling Springs (but continuing to carry private-label milk). In general these scenarios consider job and wage impacts only at Shankstead and Trickling Springs, assuming that employment at MOM's is only marginally tied to milk in this supply chain.

Because Shankstead is Trickling Springs' primary milk source, Trickling Springs would have to find an alternate organic milk supplier if Shankstead were to discontinue operations. The supply chain would likely remain intact, although with slightly different characteristics. Trickling Springs' relationship with LOFCO could facilitate this transition, as Trickling Springs is able to purchase milk from LOFCO if needed. Employment at Trickling Springs would likely be unaffected as operations from processor to the retail store would continue unchanged. Some of the three employees and $120,000 in annual wages at Shankstead would likely be affected. However, given that Shankstead has other farm operations (including chickens), it is unlikely that Shankstead would reduce its labor demand to zero. Under this scenario, the total direct economic impact would range from zero to at most three jobs, and from $0 to at most $120,000 in wages.

A closure of Trickling Springs' milk processing business would likely have the largest direct economic impact on the supply chain. Because milk processing is the main line of business for Trickling Springs, it is unlikely that the business would remain in operation. Up to 30 full-time-equivalent (FTE) jobs and $780,000 in wages could be lost in this scenario. If the Trickling Springs store, which also sells other food and non-food items in addition to milk and creamery products, was to remain open, the job losses could be reduced by between five and ten FTEs and between $130,000 and $260,000. Shankstead and its workforce would likely be unaffected because it is able to sell its milk to LOFCO under the same terms it sells to Trickling Springs (at least in the short term).

If MOM's discontinued purchases of Trickling Springs milk, some job and wage losses might occur at Trickling Springs. MOM's is Trickling Springs' largest private-label milk buyer, and fluid milk sales at MOM's account for about 17 percent of the milk that Trickling Springs processes. Assuming that Trickling Springs could not immediately make up for the lost demand, it is possible that some employment would be lost due to reduced production and fewer deliveries. If employment is reduced in proportion to reduced milk demand, this scenario could result in a loss of about five FTEs and $130,000 in wages. However, some of these losses might be offset by employment gains at other milk processors in the region if MOM's were to purchase private-label milk from another local dairy. Because Shankstead can sell excess milk to LOFCO if necessary, it is not likely this scenario will affect the farm's operation in the short term.

The supply chain may also support community linkages that are not well expressed in terms of wages, employment, or revenue. Informal community and civic relationships are more common than formal associations in this supply chain. For example, after conversion to organic production, Shankstead is more connected to neighboring farms and community members through production and consumption relationships. Farms in the area tend to use each other's products as inputs and consume each other's finished products, and a nearby residential neighborhood has allowed Shankstead to foster links with area residents. Shankstead also has minimal involvement in the Franklin County Grazers and the Pennsylvania Association for Sustainable Agriculture, organizations that link farm producers in the area.

Farmer Profitability

As Trickling Springs' sole source of raw milk, Shankstead is a key actor in this supply chain. Shankstead also relies heavily on the availability of Trickling Springs as an outlet for its milk and as a revenue source. Information about all revenues, costs, and debt service are not available to calculate net profit for a particular period of time for Shankstead. However, interviews suggest that operating as an organic, grass-based dairy is profitable for Shankstead in the long term.

Shankstead began its transition to an organic, grass-based dairy in December 2007. Prior to this Shankstead relied primarily on a corn-based grain ration for feed. The farm's fixed costs and many variable costs (e.g., labor) remained the same after transition. Comparing operating profitability under current operations to operations prior to transition, three primary factors appear to be important: grain feed prices, raw milk prices, and milk production per cow.

Shankstead began transition as organic corn prices were about to peak in 2008 and 2009. The switch to a grass-based diet resulted in a reduction in feed costs estimated at between 67 percent and 93 percent.[14] Milk production also decreased with the switch to grass-based feed. Current production at Shankstead is about 40 pounds per cow per day, which is between 25 percent and 35 percent lower than on the grain ration. Raw milk prices are higher under organic and grass-based production. Research suggests that typical prices for organic milk are on average between 20 and 33 percent higher than prices of conventionally produced milk (McBride and Greene 2009). At the time of the study, Shankstead was receiving prices for raw organic milk well above this average premium.

In summary, Shankstead's milk operations are likely more profitable after transition to organic and grass-based production, at least on a per unit basis. Although production per cow is lower with the heavier reliance on pasture for feed, reductions in grain feed costs and increased raw milk prices appear to offset production losses. Not considered in this analysis is debt service for debt related to the upfront costs of organic transition, potential changes in unpaid farm labor, and costs associated with inspections and permitting for retail raw milk sales (about 10 percent of Shankstead's production).

Prospects for Expansion

The Trickling Springs–MOM's supply chain could potentially expand at any point in the supply chain through expanded production at Shankstead, increased processing and wholesale activity at Trickling Springs, or increased demand for Trickling Springs products at MOM's. Interviews suggest that the most likely case for expansion would be an increase in business at Trickling Springs. However, such an expansion would be constrained in the short term by refrigerated storage and wholesale delivery capacity. Although Trickling Springs could handle an increase in production with its current equipment, the existing refrigerated cooler is not large enough to handle a large increase in inventory and the assembly and movement of more and larger orders.

Trickling Springs could increase its production by about 20 percent and still exclusively use milk from Shankstead. Trickling Springs currently has a surplus of milk they send to LOFCO, and it is possible that Shankstead could increase production marginally without cutting into their raw milk business. Beyond this increase, Trickling Springs would need to identify additional milk supplies. Trickling Springs could source additional milk from LOFCO or work directly with other farms in the area that produce organic milk.

Shankstead has plans to expand its operation, possibly through growth in its poultry enterprise. The existing pasture rotation, with movable poultry pens that follow grazing in the pasture about three days after the cow herd, ideally would support at least twice the size of the current poultry flock. Some plans are in place to improve pasture efficiency and further reduce grain feed requirements for the herd, but this would not require an increase in the cow herd or necessarily increase milk production. At most, Shankstead could increase milk production by 10 percent if necessary.

Policy and Regulatory Issues

As with other milk supply chains, a relevant policy issue is inspections and quality standards for milk production. In this case Shankstead and Trickling Springs invest considerable time and energy in ensuring that their product meets state and federal regulations. Both Shankstead and Trickling Springs engage in inspections and testing that exceed legal requirements for typical milk marketing. Some of these additional practices are due to the requirements of particular wholesale customers (although not MOM's at the time of this study) that conduct their own inspections at the farm and in processing stages. Both Shankstead and Trickling Springs are required to document their certification as an organic producer and handler, respectively. In addition to organic certification, Shankstead is also subject to more stringent regulations in order to sell raw milk in Pennsylvania. Neither Shankstead nor Trickling Springs has experienced problems with product recalls or had problems meeting quality standards for milk in the supply chains studied.

Trickling Springs is also subject to laws and regulations governing the trucking industry. Drivers are limited by federal motor carrier regulations on the amount of time they can spend on the road over a given time period, which limits the number of delivery stops they can make on each route. Trickling Springs has not had difficulty meeting these regulations, but they could pose logistical challenges if expansion of Trickling Springs' business necessitates additional delivery stops and routes.

Key Lessons

The intermediated supply chain studied in this case involves fluid milk products that are locally produced and sold in a small chain of retail grocery stores in the DC area. Three key lessons emerge from this study. First, quality and production practices initiated at the farm level are important to the retailer and (presumably) the customer. Private-label milk brands in larger supermarkets typically convey few production characteristics beyond meeting quality standards and in some cases organic and rBST hormone-free production practices. In this case, locality of the source matters to some degree for MOM's. Shankstead clearly tries to differentiate itself in terms of quality and production practices and to communicate those characteristics to customers through Trickling Springs. The Trickling Springs name is retained on the MOM's private-label milk packaging, but MOM's does not actively promote the locality and origin of its private label at the retail point of sale or in its flyers.

A second lesson is that the supply chain largely relies on the close working relationship between Shankstead and Trickling Springs. Trickling Springs currently sources all its milk from Shankstead, and Trickling Springs is the sole processor of Shankstead milk. Although both are somewhat insulated against a breakdown in this relationship (because of the existence of LOFCO), the current arrangement is clearly mutually beneficial. For example, Trickling Springs is able to identify a single source of their milk for customers (like MOM's) who demand particular milk characteristics. Also, Shankstead is able to receive a premium for its investments in production practices and deal with a single buyer who values their organic and grass-based practices.

Finally, it is important to note that the local and quality differentiated product in this supply chain is filling the role of a grocery staple at the MOM's stores. Trickling Springs milk sold under the MOM's private label appears to be a loss leader for MOM's, rather than a niche product carried to satisfy a minority of its customers (like, for example, Trickling Springs milk sold in glass bottles at MOM's). The MOM's chain may itself represent a niche in the larger grocery and supermarket industry in Washington DC, but its private-label milk is filling a role similar to other supermarket private-label brands.

Cross-Case Comparisons

The three supply chain cases for milk in the Washington DC area—Maryland and Virginia Milk Producers Co-op, South Mountain Creamery, and Trickling Springs Creamery—demonstrate the variety of ways that milk can move from producers to consumers. This variety is especially intriguing given the relative homogeneity of milk as a product compared to some other grocery staples. The mainstream supply chain studied here represents a common way the majority of consumers satisfy their product demands for milk, while local supply chains illustrate unique methods for producing, processing, and marketing milk that are filling a viable (if small) market niche.

Supply Chain Structure

Supply chain structure varies greatly among the three cases presented above. Structure largely reflects differences between supply chains in market position and product and service characteristics offered by enterprises in the supply chains.

Differences in structure may be related to information conveyed to consumers about where, how, and by whom the product was produced. In the direct market supply chain, South Mountain labels identify the geographic origin and farm name where the milk is produced and processed. Trickling Springs identifies the location of the milk processor on product labels and that the milk is organically produced. As of October 2009 the Trickling Springs web site mentioned that its milk was sourced from nearby farms but did not specifically identify Shankstead as the supplier. On the private-label products from Trickling Springs, the processor location is less prominently displayed. In the mainstream case only the location of the processor or distributor is identified on private-label products sold by Maryland and Virginia Co-op.

Durable relationships are evident in the mainstream and intermediated supply chains, both of which involve multiple supply chain relationships that are critical to success. Interdependence, trust, and information sharing have likely developed between the cooperative (Maryland and Virginia Co-op) and its private-label customers. Consequently, efficiently managing a large volume of milk for many customers and stores requires a high degree of coordination and communication. In the intermediated chain, Shankstead is an important supplier for Trickling Springs, and Trickling Springs values the unique production and product characteristics maintained by Shankstead.

Collective organizations play a prominent role in the mainstream supply chain but a minimal role in the intermediated and the direct marketing chains. The mainstream producer cooperative (Maryland and Virginia Co-op) is a key enterprise in the supply chain for private-label milk in the DC area, responsible for production, processing, and distribution. Trickling Springs sells milk to consumer retail cooperatives, although these are not Trickling Springs' largest accounts. The Lancaster Organic Farmers Cooperative (LOFCO) plays a small but important role in the intermediated supply chain. LOFCO acts as a balancer for Trickling Springs, buying excess milk when Trickling Springs cannot process all of Shankstead's production and selling organic milk to Trickling Springs when demand increases.

Supply Chain Size and Growth

The direct market and intermediated supply chains are relatively small compared with the mainstream chain. The two local supply chains combined handle only a small fraction of the milk produced and distributed in the mainstream chain. This general pattern likely holds for the DC-area milk market as a whole.

Growth for the milk supply chains studied is likely to be driven largely by broad trends in consumer preferences for milk relative to other beverages. This is the case especially for the mainstream supply chain, which relies exclusively on an undifferentiated product through supermarkets. Aside from a general increase in the demand for milk, the Maryland and Virginia Co-op can likely grow only by capturing a larger share of a fixed (or possibly declining) market for fluid milk.

The local supply chains, by virtue of occupying smaller niches within the market for fluid milk, could conceivably grow if consumer demand for certain product and service characteristics (e.g., organic production, glass bottles, and home delivery) increased. The processors in both local chains have developed their own processing and distribution capacity and are capable of accommodating some short-term growth. In general, growth among local supply chains for milk would likely occur through internal expansion of existing firms, rather than new market entrants. This is due primarily to the large fixed costs required to process and distribute fluid milk. However, expansion of processing capacity is also expensive, and it is not a foregone conclusion that Trickling Springs or South Mountain could easily grow if demand outstripped their existing capacity.

Current operations or potential future growth does not appear to be hampered by regulatory compliance. Fluid milk is a highly regulated and monitored industry requiring inspections and compliance at multiple stages of the supply chain. Some retailers also require additional inspections beyond state requirements. Aside from some initial difficulties as the local supply chains began processing operations, it does not appear that they have more difficulty than mainstream chains in meeting these requirements. However, achieving compliance may be costly for new enterprises; South Mountain initially found it difficult to work with state regulators to identify and resolve compliance problems.

Supply Chain Performance

Differences in structure and size naturally give rise to differences in performance among the milk supply chains. Retail prices, the distribution of revenues and costs among supply chain participants, contributions to local economies, and transportation fuel use are all determined by how supply chains move product from farms to consumers and the market position of each supply chain.

Producers receive a greater share of retail revenue in the direct and intermediated supply chains (table 20). South Mountain retains 100 percent of the revenue (because they sell directly to consumers), but they also incur processing and marketing costs totaling an estimated 63 percent of the retail revenue. Shankstead receives about 39 percent of the retail revenue in the intermediated supply chain, compared to 32 percent for farms in the mainstream supply chain. Farm revenues per unit for milk, net of marketing costs, are significantly higher in the local supply chains, although there is little difference in producer revenue per unit between the direct market and intermediated supply chains.

Differentiation beyond "local" is necessary to receive price premiums for milk. The local supply chains are differentiated by production characteristics (organic and grass-based production for Shankstead), service (home-delivery for South Mountain), and packaging (glass bottles). Private-label milk in supermarkets is consistently priced lower than the other milk products in the marketplace. Branded organic milk (i.e., not local) is priced higher than the local products, even when milk is both local and organic (e.g., Trickling Springs), suggesting that the local attribute does not garner any additional retail price premium.

Table 20. Allocation of retail revenue in DC area—milk supply chains, by supply chains and segment

	MAINSTREAM		DIRECT		INTERMEDIATED	
	Maryland and Virginia Co-op[a]		South Mountain Creamery		Trickling Springs—MOM's[b]	
Supply chain segment	Revenue ($/half gal)	% of total	Revenue ($/half gal)	% of total	Revenue ($/half gal)	% of total
Producer(s)[c]	0.64	32.3	1.22	37.5	1.29	39.2
Producer estimated marketing costs[d]	na	na	2.03	62.5	na	na
Dairy cooperative[e]	0.18	9.0	na	na	na	na
Processor[f]	0.58	28.9	na	na	1.82	55.3
Retailer	0.59	29.8	na	na	0.18	5.5
Total retail value[g]	1.99	100	3.25	100	3.29	100

[a.] Mainstream chain revenue allocations are calculated from the Virginia State Milk Commission Presumed Costs reports, Eastern Market, for plastic half-gallon 100+ cases. Estimates are based on 3-month averages, September—November 2009. These reports do not specifically identify revenue allocations for the Maryland and Virginia Cooperative or its retail customers and are representative of the milk industry in the DC area in general.

[b.] Revenue shares calculated for Trickling Springs milk sold as MOM's private-label milk. Trickling Springs—labeled glass bottles add $0.30 per half gallon to the retail value, which accrues solely to the retail stores.

[c.] Mainstream: Based on September—November 2009 3-month average class 1 price announcement for Federal Milk Order Number 1, Frederick MD/New Holland PA ($14.95/cwt). Direct: The dairy farm also operates as the processor.

[d.] Includes the estimated portion of producer revenue attributed to costs of processing and home delivery, including labor costs. Total per unit revenue for the producer is 1.22 + 2.03 = 3.25 ($/half gal).

[e.] Calculated as the difference between raw product costs in the VA Presumed Costs reports and the class 1 price announcement (i.e., producer revenue). Includes revenue that may accrue to the cooperative or third party milk haulers.

f. Mainstream: Calculated as the difference between wholesale delivered costs and raw product costs from the Virginia State Milk Commission Presumed Costs reports. Includes revenues attributable to delivery to the retail stores. Intermediated: Trickling Springs operates as both the processor and distributor to retail stores.

g. Mainstream: Median retail price of half gallons, January—December 2009. Direct: Half-gallon prices listed on the South Mountain web site as of December 2009. Intermediated: Median retail price of half gallons, January—December 2009.

Source: King et al. (2010: 50).

Wages and business proprietor income for all supply chains accrue primarily within the DC area. All wages and income in the direct market and intermediated chains accrue within the DC area. In the mainstream chain, corporate ownership of large supermarket chains may be based outside the region, but the Maryland and Virginia Co-op headquarters, many of the dairy farms, the processing plants, and retail stores are located within the DC area.

The distance the product travels from production to consumers (food miles) is 48 miles in the direct supply chain and 94 miles in the intermediated supply chain (table 21). However, the intermediated supply chain uses less fuel per unit of product delivered when compared with the direct market supply chain. This is due to larger load sizes making fewer delivery stops for Trickling Springs. Although food travels about twice as far in the intermediated supply chain, trucks on Trickling Springs' delivery routes to MOM's carry more than four times as much product.[15]

Key Lessons

The case studies of milk supply chains in the Washington DC area illustrate the following three general lessons.

1. Direct and intermediated supply chains for milk currently capture a relatively small portion of the total market for milk in the DC area, but they fill a market niche where consumers are willing to pay extra for certain product and service characteristics. These supply chains appear to rely on differentiation to receive a premium over mainstream milk products (prices are about 64 percent higher in

Table 21. Food miles and transportation fuel use in DC area—milk supply chains

SUPPLY CHAIN SEGMENT[a]	FOOD MILES[b]	TRUCK MILES
Direct: South Mountain Creamery		
Home delivery[d]	48	175
Intermediated: Trickling Springs Creamery		
Farm to processing plant	9	18
Processing plant to retail Stores[e]	85	250
All segments	94	

Milk volumes expressed in hundredweight (cwt); one hundredweight of milk is equal to approximately 11.6 gallons.

[a.] Food miles, fuel use, and product volume in the mainstream supply chain were not available.

[b.] South Mountain: Distance calculated from South Mountain to the Maryland-DC border at Chevy Chase Circle. Trickling Springs: Plant to retail stores segment calculated as average distance to the five MOM's stores.

[c.] Fuel use for Trickling Springs is in gallons of diesel fuel; South Mountain fuel use reported as gallons of gasoline.

[d.] Delivery routes also carry nonmilk products. Fuel use is calculated as the milk portion of total fuel use based on the average share of each load that is accounted for by milk (about 90 percent).

[e.] Delivery routes that serve MOM's stores also serve other accounts. Fuel use is apportioned to the MOM's deliveries based on the average share of each load that is accounted for by MOM's milk deliveries (about 36 percent).

Source: King et al. (2010: 51).

RETAIL WEIGHT (CWT)	FUEL USE (GAL)[c]	FUEL USE PER CWT SHIPPED
9.2	17.5	1.90
160	3.6	0.02
41.1	31.1	0.76
		0.78

the local supply chains) and on diversification of products and enterprises to maintain multiple revenue streams.

2. Differentiation and diversification may be a response to relatively high per unit processing and distribution costs. For the direct and intermediated supply chains, offering a variety of products through several market outlets allows Trickling Springs and South Mountain to increase revenue per unit of milk delivered to customers.

3. Large economies of scale keep processing and distribution costs in the mainstream supply chain well below those in the local supply chains, necessitating product differentiation to maintain price premiums. For example, processing costs per unit in the mainstream chain may be less than half of the costs per unit in the direct marketing chain.[16] On the other hand, the high fixed costs necessary to process and distribute milk may prevent entry of possible competitors into other local supply chains for milk. This barrier to entry may help preserve the market niche for smaller enterprises and prevent the erosion of price premiums for differentiated products and services.

4. Although the products in the two local supply chains are distinguished by their origin, the locality of production and processing is not used as a primary differentiating characteristic. Neither does the cooperative in the mainstream case attempt to identify its product as local.

5. Product labels in the local supply chains identify where the product comes from, but only in the direct market case is the farm identified. Milk sold in the mainstream case is processed and primarily sourced from within the DC area, although it is not marketed with any designation of origin or identification of the producer. Thus a large portion of the milk sold in the DC area meets the definition of a local product, but the lack of information about the milk's origin means that it is not marketed through a local food supply chain under the definitions used in this study.

Notes

1. Despite the abundance of markets in the area, only a small amount of fluid milk is marketed through farmers markets.
2. Herd size includes both milking and dry cows; Calvert County, Prince George's County, Fairfax County (all in Maryland), and Rappahannock County, Virginia, were not reported due to disclosure issues. County-level milk production estimated based on each county's share of the state total herd size.
3. Precise information about the distance traveled from farms to the processing plants, and from plants to retail stores, was either not available or could not be disclosed due to confidentiality concerns.
4. September–November three-month average class 1 price announcement for Federal Milk Order Number 1, Frederick MD/New Holland PA ($14.95/cwt). One hundred pounds of milk equals about 23.26 half gallons. A larger total volume of milk is sold in gallon containers at larger supermarket chains. This study bases price comparisons and other analyses on the price of half gallons because gallon containers are less common in the other supply chains.
5. Production costs were not available for Maryland and Virginia Co-op member farms. Average milk production costs for 2009 were $14.74/cwt in the Northern Crescent production region (which includes most of Pennsylvania and Maryland), and $15.36/cwt in the Southern Seaboard region (which includes most of Virginia, Delaware, and parts of Maryland). See "Commodity Costs and Returns: Data," available at ers.usda.gov/data/costsandreturns/testpick.htm (accessed 6/16/2010). Farms with smaller herds tend to have higher production costs per hundredweight (MacDonald et al. 2007).
6. Price data were collected for whole milk in half-gallon containers during 2009 through informal in-store observations at two supermarket chain locations.
7. Prices are listed on the South Mountain web site. This was the price listed in November 2009.
8. Fuel costs and driver wages calculated based on total full-time drivers and delivery route driving distances reported in interviews. Vehicle costs are calculated from

per-mile heavy-duty truck cost estimates for tires ($0.04 per mile), depreciation ($0.09 per mile), and maintenance and repair ($0.12 per mile), adjusting for inflation (Barnes and Langworthy 2003). Calculations based on 430 hundredweight of milk sold per week.

9. The exact geographic distribution of South Mountain's customers, which would allow for calculations of average or median distance from farm to consumer, is not known. Food miles are calculated as the driving distance from South Mountain to the Maryland–Washington DC border at Chevy Chase Circle.

10. The size of the gross margin under cooperative marketing depends on the assumed price of milk received. At the time of the interviews, milk prices were around $13/cwt, which is low compared to recent averages (prices were about $18/cwt in January 2010).

11. Milk accounts for about 90 percent of the volume and weight of deliveries. To calculate gross marketing margins for milk under home delivery, 90 percent of delivery and other nonprocessing costs were apportioned to the milk operation. Delivery fee revenues were apportioned based on the revenue share of milk (37 percent).

12. To calculate the wage bill, the midpoint of the hourly wage range—$13/hr—was used for all 30 workers. Full-time workers are assumed to work 2,000 hours per year.

13. To calculate precisely the fuel use that is due to MOM's orders, we would need to be able to calculate fuel use for the counterfactual case; that is, without deliveries of fluid milk to MOM's. It is unknown how the supply chain logistics (e.g., the number of delivery routes and distance they travel) would change in that case, although it is likely that Trickling Springs would use fewer delivery routes rather than less product loaded on the same number of routes.

14. Estimated change in feed costs calculated using September 2009 prices, reported to be about $8/bu. Reported grain requirements under the grass-based system are between two and 10 pounds per cow per day.

15. Information about food miles and fuel use was not available for the mainstream supply chain. Also, the reported food miles and transportation fuel use calculations do not account for differences in consumer transportation to and from retail stores, which is a part of the mainstream and intermediated supply chains (but not the direct market supply chain in this case).

16. Reported processing costs for South Mountain were about $1.00 per half gallon. Processing costs in the mainstream chain are estimated using the difference between the three-month average platform costs and raw product costs from the Virginia State Milk Commission Presumed Costs reports, Eastern Market (Virginia Department of Agriculture and Consumer Services 2009), which was approximately $0.44 per half gallon.

References

Barnes, G., and P. Langworthy. 2003. "The Per-Mile Costs of Operating Automobiles and Trucks." Technical report #2003–19. St. Paul: Minnesota Department of Transportation.

Bonanno, A., and R. A. Lopez. 2005. "Private Label Expansion and Supermarket Milk Prices." *Journal of Agricultural and Food Industrial Organization* 3(1), Article 2. bepress.com/jafio/vol3/issl/art2/ (accessed 12/28/2011).

California Department of Food and Agriculture. N.d. "Retail Prices: Nielsen Retail Pricing." cdfa.ca.gov/dairy/retail_prices_main.html (accessed 11/19/2009).

Dudlicek, J. 2009. "Corporate Profile: Maryland and Virginia Milk Producers." *Dairy Foods* 110(1). dairyfoods.com/Archives?issue=1864251 (accessed 5/25/2010).

King, Robert P., Michael S. Hand, Gigi DiGiacomo, Kate Clancy, Miguel I. Gómez, Shermain D. Hardesty, Larry Lev, and Edward W. McLaughlin. 2010. *Comparing the Structure, Size, and Performance of Local and Mainstream Food Supply Chains.* ERR-99. Washington DC: U.S. Department of Agriculture, Economic Research Service.

McBride, W., and C. Greene. 2009. *Characteristics, Costs, and Issues for Organic Dairy Farming.* ERR-82. Washington DC: U.S. Department of Agriculture, Economic Research Service.

MacDonald, J. M., E. J. O'Donoghue, W. D. McBride, R. F. Nehring, C. L. Sandretto, and R. Mosheim. 2007. *Profits, Costs, and the Changing Structure of Dairy Farming.* ERR-47. Washington DC: U.S. Department of Agriculture, Economic Research Service.

U.S. Census Bureau. 2009. *Population Estimates, County Totals: Vintage 2009.* census.gov /popest/data/counties/totals/2009/index.html (accessed 12/28/2011).

USDA-AMS. N.d. "Farmers Market Search." Washington DC: U.S. Department of Agriculture, Agricultural Marketing Service. search.ams.usda.gov/farmersmarkets/ (accessed 11/23/2009).

USDA-ERS. N.d. "Food Availability Spreadsheets: Dairy (Fluid Milk and Cream)." Washington DC: U.S. Department of Agriculture, Economic Research Service. ers.usda .gov/data/foodconsumption/FoodAvailSpreadsheets.htm#dyfluid (accessed 5/10/2010).

USDA-NASS. 2009. *Census of Agriculture, 2007.* Washington DC: U.S. Department of Agriculture, National Agricultural Statistics Service.

———. 2010. "Milk Production, Disposition, and Income: 2009 Summary." Washington DC: U.S. Department of Agriculture, National Agricultural Statistics Service.

Virginia Department of Agriculture and Consumer Services. 2009. "Presumed Cost Schedule." September–November. vdacs.virginia.gov/smc/publications.shtml (accessed 10/17/2009).

Part 3

A Synthesis of Case Study Findings

8 Product Prices and Availability

Kristen S. Park, Miguel I. Gómez,
Gerald F. Ortmann, and Jeffrey Horwich

Two important questions for members of supply chains that produce and distribute local food products are: "When are local products available?" and "Does the attribute 'local' exhibit retail price premiums in the marketplace?" In this chapter we address these two critical characteristics of products sold through local supply chains.

Methods

Parallel to this series of coordinated case studies, we collected data on weekly availability of products, prices, varieties, attributes (e.g. organic, local) and package types of the five case study products: apples, blueberries, spring mix, ground beef, and fluid milk. The weekly data were collected from a variety of retail outlets in the five different study regions over the course of one year. Retail outlets in each study region generally included two supermarket chains, two natural foods stores, and two farmers markets. Although these three retail outlets are typical for these products, they may not be fully representative of all potential retail outlets through which both local and nonlocal products are sold. Data on product prices and product availability were collected for 51 weeks from January 10, 2009, through December 26, 2009.

A systematic analysis of the retail prices collected in this study that controls for multiple product attributes can provide valuable information on price premiums for local food products. For the purposes of this exercise, a local product was defined as one that is raised, produced, and processed in the locality or region where the final product is marketed.

For a product to be considered local, the label or marketing materials had to convey information about where it was produced. For example, store brand milk was defined as being domestically produced in the United States but not as being local. Even though in most cases the milk was produced and processed within the local geography, it did not convey information about where it was produced.

Product Availability

It has been argued that seasonality in production (in the case of perishable products such as fruits and vegetables) and lack of a local infrastructure (in the case of processed foods such as dairy and meats) are significant barriers to increasing the share of local foods in the overall food system. In addition, we know very little about the extent to which local products are delivered through mainstream supply chains. Here we present findings about the availability of local foods, comparing mainstream, direct market, and intermediated supply chains.

Apples

In general, apples were available nearly year-round in supermarkets and natural foods stores in the five case study regions. Although apples are a crop harvested in the fall, they can be stored for relatively long periods in coolers and even longer in specialized controlled atmosphere coolers. They can also be shipped over long distances and can be imported should domestic stocks or special varieties run low.

Figure 25 illustrates the availability of local apples in the five study locations and in all the market channels. Large apple production regions, such as New York and Oregon, carry local apples all year in supermarkets and farmers markets, having plentiful supplies of local product. Natural foods stores, however, tend to carry organic apples, and organic apple production is somewhat difficult, especially in New York, the District of Columbia, and the Twin Cities study regions. Sources of local organic supplies are limited in these regions, although natural foods stores usually carry local apples during harvest and winter months.

Local apples were almost always available in farmers markets when markets were open. The exception was in the Twin Cities area, where local

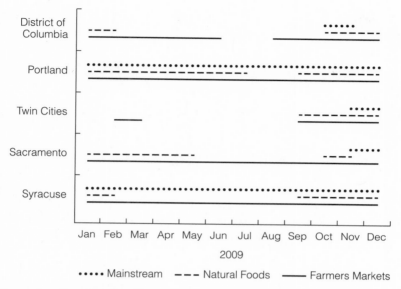

Fig. 25. Local apple availability across study locations and retail outlets. Source: Weekly data collected by authors.

apples were not available in most of the late spring and summer months, and in the District of Columbia for a portion of the summer. One reason why local apples might be plentiful compared to other commodities is because they can store well, especially under controlled atmosphere storage and/or with certain varieties.

Blueberries

Blueberries from all sources were available almost year-round in at least one supermarket (mainstream) channel in three of the five locations (Sacramento, Twin Cities, and Syracuse). October was the exception, when blueberries, whether local or not, were frequently unavailable until imported supplies become available in November.

Figure 26 presents data on availability of local blueberries in retail outlets across the five metropolitan statistical area (MSA) study locations. Local blueberries were available seasonally in all five MSA study locations. Availability of local blueberries varied from 16 weeks in Sacramento to only five weeks in the Twin Cities. In all five study locations, local blueberries

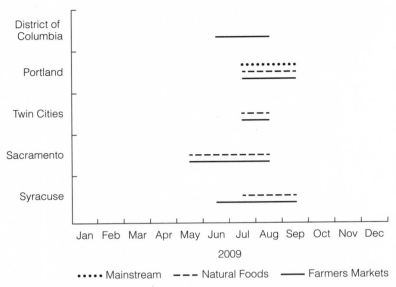

Fig. 26. Local blueberry availability across study locations and retail outlets.
Source: Weekly data collected by authors.

were available through farmers markets. The maximum number of market vendors selling blueberries varied from just one at each Minneapolis market and Syracuse market to 15 at one Portland farmers market.

Of the 11 supermarkets studied, only two (both supermarkets in Portland) sold any local blueberries. Portland is the only study location near a commercial blueberry production area, which explains why local blueberries are of a volume sufficient to enter into mainstream distribution.

Four of the five study locations (the District of Columbia area was the exception) sold local berries through natural foods stores as well as through farmers markets. At retail, all blueberries are sold in packages rather than in bulk. The industry recognizes the importance of encouraging larger retail pack sizes as a driver for selling more product (Offner 2009).[1] In farmers markets, especially in Portland, discounts are offered for flats and half flats of blueberries (12 pints and six pints respectively). In mainstream distribution, larger pack sizes have become increasingly prevalent in some markets. In Portland, packs of one pound or more (maximum size of five pounds) were available for 26 weeks in 2009.

Similarly, there were 24 weeks in Syracuse when blueberries were available in pack sizes of one pound or more (maximum size of two pounds). Neither the Twin Cities nor the District of Columbia area reported any pack sizes of one pound or more.[2]

Organic blueberries were widely available through supermarkets and natural foods stores in all five locations, although not as consistently as conventional blueberries. In contrast, four farmers markets (those in Syracuse and the Twin Cities) did not have any organic blueberries over the course of the season. The two Portland farmers markets featured multiple organic blueberry vendors (as many as five on a given day), while the District of Columbia area market and the two Sacramento markets never had more than a single organic blueberry vendor.

Spring Mix

Spring mix in general was available consistently at supermarkets and natural foods stores in all locations except Syracuse, where the natural foods store did not carry any spring mix during two periods lasting at least 10 weeks. With the exception of farmers markets, spring mix is sold primarily in packages, packed usually in four- and five-ounce bags, and five-ounce and one-pound clamshells. Unlike virtually every other produce item in the United States food system, the number of organic products available in a geographic market in this study is equal to or greater than that for conventionally grown product. In most markets, conventionally grown spring mix is available solely in five-ounce packages. The natural foods stores in this study marketed organic spring mix only.

Local spring mix availability across the five MSA study locations in mainstream and direct channels is presented in figure 27. The availability of local spring mix was relatively limited; only farmers markets in Sacramento, the Twin Cities, District of Columbia, and Portland, along with natural foods stores in the Twin Cities and Portland, carried local product for extended periods. No local spring mix was observed in mainstream retail outlets. Moreover, seasonality does not appear to preclude the supply of local spring mix in four of the case study areas. In the District of Columbia some vendors use greenhouses along with cold-weather tolerant greens in the "spring" mix to extend their season. Local spring mix was not available in any of the study outlets in the Syracuse MSA.

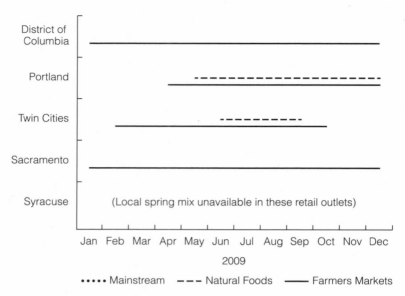

Fig. 27. Local spring mix availability across study locations and retail outlets.
Source: Weekly data collected by authors.

Ground Beef

Figure 28 shows data on availability of local ground beef across the five study locations in each supply chain. Seasonality is not an issue with the availability of local ground beef because the product is usually sold frozen. The Twin Cities consumers had year-round access to ground beef from local food supply chains at nearly all observed retail outlets, including farmers markets, natural foods stores, and mainstream retailers. This is in marked contrast to most other study locations, where ground beef from local food supply chains was available only at one farmers market and, in Syracuse and in Portland, at one natural foods store.

Generally there is greater product differentiation among nonlocal food supply chains with more products offered across a larger number of retail channels. The Twin Cities is the only exception, where six different labels for ground beef products were available through local food supply chains. It is noteworthy that beef products from local food supply chains tend to fill unique product attribute and retail channel niches. Beef products from local food supply chains do not appear to be in

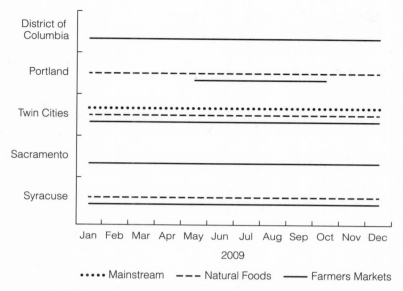

Fig. 28. Local ground beef availability across study locations and retail outlets. Source: Weekly data collected by authors.

direct competition with products from nonlocal food supply chains at the same retail location.

Milk

The availability of local fluid milk across the five MSA study locations is illustrated in figure 29.[3] Local milk was available nearly year-round at many retail outlets in all study areas except Sacramento. The Twin Cities study area had the greatest penetration of local milk, which was available across all retail outlets—mainstream, natural foods stores, and farmers markets—for the full study year. In this study area several Minnesota-based dairies sold branded milk (which clearly denotes the product's origin) in a variety of market outlets. Local milk in Syracuse and Portland was available in at least one of the observed farmers markets, at least one natural foods store, and at least one supermarket.

In the District of Columbia, local milk was available at the farmers market and natural foods stores only. Mainstream supermarkets in these study areas tend to carry private-label milk that does not reveal the product's origin.

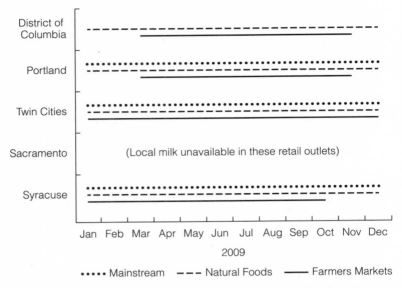

Fig. 29. Local milk availability across study locations and retail outlets. Source: Weekly data collected by authors.

Private-label milk sold in supermarkets may be locally sourced, but it is not identified as such on product labels or other point-of-sale information.

Summary of Local Food Availability

Availability of local product appears to depend on local production, market demand, and ability to store the product. The scope of available product or the depth of market penetration seems likely to depend on the quantity of local product and size of individual producers able to supply many different retail outlet types, whether supermarket, natural foods store, or farmers market. The length of season during which local product is available depends on the product itself, including its growing season and/or storage capability.

Prices

There is very limited evidence in the literature on price differentials between locally produced foods and other sources using actual retail price data. Most studies on prices for local foods elicit consumer willingness to pay for local foods but measure consumer intentions and not actual consumer behavior. Eastwood, Brooker, and Orr (1987) found no local preference except in the case of tomatoes in Tennessee. They postulated that perhaps there were regional or geographic differences in demand for local products and/or that a preference for local was an emerging trend.

Various studies since then have found willingness to pay a premium for local products, which varies by geography, product, and consumer demographics. Loureiro and Hine (2002) found that consumers in Colorado were willing to pay approximately 9 percent more for local potatoes. Brown (2003), in Missouri, found 58 percent of consumers unwilling to pay a premium for any local foods, but 22 percent were willing to pay at least a 5 percent premium. Darby and colleagues (2008) found a willingness to pay a premium for locally produced strawberries in Ohio of $0.64/lb and $1.17/lb for supermarket and direct market shoppers respectively (averaging 27 percent of the retail price).

Giraud, Bond, and Bond (2005) found consumer preferences for locally made specialty food products across northern New England and a willingness to pay a premium averaging 9 percent. This willingness to pay varied by the base price of the product but was similar across three New England states. The researchers therefore argued that willingness to pay varies by state and by product as well as demographics. Carpio and Isengildina-Massa (2009) conducted a phone survey of South Carolinian consumers and found that consumers are willing to pay price premiums for locally grown products. Respondents indicated willingness to pay premiums of 23 percent and 27.5 percent for animal products and produce respectively. Hinson and Bruchhaus (2005) found Louisiana consumers have an average willingness to pay a premium of 21 percent for strawberries.

Regression techniques were used to develop five models (one each for blueberries, spring mix, 2 percent milk, ground beef, and apples) in which the dependent variable is the price per unit of volume of the product. In the models, product price is a function of a vector of product attributes (local, organic, conventional, package size, and price promotion), metropolitan area, retail outlet type, and season of the year. Prices were aggregated by retail type within each metropolitan area, and the median aggregated price was used to control for differences in variability in pricing between retailers of the same type. The variables employed in the analysis and their definitions are listed in table 22.

Results

Table 23 summarizes the results of the econometric model explaining retail prices for the five focal products (detailed results are presented in table 26). In table 23 the positive and negative signs indicate a direct and inverse relationship between the given variable and the retail price, respectively. The results indicate that a price premium for local product exists for blueberries, spring mix, ground beef, and 2 percent fluid milk but not for apples. Results indicate that many, if not most, product attributes—including origin, organic production, and package size, among others—are in general significant determinants of retail prices. In the case of ground beef, the labels "local," "percent lean," "organic," "natural," and "grass-fed" carry significant price premiums. In the case of apples the attributes "organic," "type of packaging," and "variety" illustrate other important product attributes affecting retail prices. Many non-product-attribute factors also affect prices, including seasonality, geographic regions, and retail outlet type. Price promotions, by definition, are price reductions, and these are negative and significant for every product. We first discuss the estimated price premiums associated to the attribute "local" and, in turn, explain other factors determining retail prices for each product.

Estimated Price Premiums for the "Local" Attribute

Price premiums as a percentage of the average retail price for the most common package sizes are calculated and shown in table 24. Premiums for the attribute "local" were estimated as 8.7 percent for blueberries, 20.8 percent for spring mix, 30.9 percent for ground beef, and 16.2 percent for 2 percent milk. Direct comparisons of these premiums to the premiums in the willingness-to-pay studies are not straightforward because the products considered in those studies are not the same as the ones considered here. However, willingness-to-pay studies focusing on the most perishable products (Louisiana and Ohio strawberries, Michigan greens, and Florida fresh produce) posted a range of 21 to 36 percent premiums. This is higher than the price premiums for blueberries and spring mix in the econometric analysis (8.7 percent and 20.8 percent, respectively). In

Table 22. Variables and definitions

INDEPENDENT VARIABLE	DEFINITION
Season of the year	indicated by dummies for Spring, Summer, and Fall (Winter = reference variable)
Price promotion	Sale or promoted price = 1; otherwise = 0
Package size	2% milk = ½ gallon; blueberries and spring mix = ounces; apples and ground beef = pounds
Metropolitan area	indicated by dummies for Syracuse, Washington DC, Portland, and Sacramento (Twin Cities = reference variable)
Retail type	indicated by dummies for natural foods store and farmers market (supermarket = reference variable)
Place of origin	indicated by dummies for imported and local (U.S. domestic = reference variable)
Organic	Organic = 1; otherwise = 0 (organic ground beef is included as a production claim as indicated below)

Additional variables for apples:

Bulk—no packaging	Bulk packaging = 1; otherwise = 0
Apple variety	indicated by dummies for Braeburn, Empire, Haralson, Honey Crisp, Pink Lady, Gala, and Fuji (combined), and Other; (Red Delicious = reference variable)

Additional variables for spring mix:

Bulk—no packaging	Bulk packaging = 1; otherwise = 0

Additional variables for ground beef:

% Lean	indicated by dummies for 80% lean, 90% lean, and 95% lean
Production claim	indicated by dummies for natural, grass-fed, and organic

Table 23. Relationship between retail prices and product, market, and tetail outlet attributes

	APPLES ($ PER LB)	BLUE-BERRIES ($ PER LB)	SPRING MIX ($ PER LB)	GROUND BEEF ($ PER LB)	2% MILK ($ PER 1/2 GAL)
Product Attributes					
Local	NS	+	+	+	+
Organic	+	+	+	+	+
Bulk	+	NS	NS	NS	NS
Package size	-	-	-	NS	NS
Price Promotion	-	-	-	-	-
Season					
Spring	NS	+	NS	NS	NS
Summer	+	-	NS	-	-
Fall	NS	-	NS	-	-
Metropolitan Area					
Syracuse NY	-	-	+	+	+
Washington DC	-	NS	NS	+	+
Portland OR	-	-	-	-	NS
Sacramento CA	-	+	-	+	+
Retail Outlet					
Natural foods	-	+	+	NS	NS
Farmers market	-	NS	NS	-	+

An NS means no significant relationship with retail prices.

Source: Estimates from econometric model (see table 26).

Table 24. Percent price premiums found for local and organic products

LOCAL	% PRICE PREMIUM FOR LOCAL	% PRICE PREMIUM FOR ORGANIC
Apples	(not significant)	18.0
Blueberries	8.7	27.9
Spring Mix	20.8	12.9
Ground Beef	30.9	63.5
2% Milk	16.2	82.9

Source: Estimates from econometric model (see table 26).

general the price premiums for the "local" attribute observed in this study are lower than what has been reported in recent willingness-to-pay studies.

Results of the econometric model also suggest that the attribute "organic" has a significant and positive effect on retail price for all products. Premiums for organic products were calculated as 18.0 percent for apples, 27.9 percent for blueberries, 12.9 percent for spring mix, 63.5 percent for ground beef, and 82.9 percent for 2 percent milk. These are substantially higher than the estimated price premiums for the attribute "local" (see table 24). A relevant question for private and public decision makers is whether the number of vendors in a farmers market influences the retail price level. The sidebar highlights the complexity of understanding the economics of this emerging distribution channel.

Farmers Market Pricing

JEFFREY HORWICH

Among local food supply chains, there is no more popular or public setting than the farmers market. Yet we know little, quantitatively, about how farmers markets work as an economic exchange. Specifically, how are prices determined?

We do know from vendor surveys that farmers' approaches to pricing vary greatly. Logozar and Schmit (2009) surveyed vendors in upstate New

Table 25. Farmers market pricing

Sample of regression results (selected to demonstrate range of products and outcomes)

Dependent Variables: ln(farmers market Price)—median and minimum

PRODUCT	MARKET	MAX. VENDORS
Apples (Fuji)	Sacramento1	4
Apples (Haralson)	TwinCities1	10
Apples (Empire)	Syracuse	14
Blueberries (conventional)	Portland1	10
Blueberries (organic)	Portland1	5
Spring Mix (conventional)	TwinCities1	27
Spring Mix (conventional)	TwinCities2	5
Spring Mix (organic)	Portland1	8
Ground Beef (natural)	Twin Cities1	4
Ground Beef (natural)	Syracuse	3

* = Expected (negative) result (*** = significance at 1%, ** = 5%, etc.).

! = Unexpected (positive) results (!!! = significant at 1%, !! = 5%, etc.).

( ) = No statistically significant relationship.

York and found "cost plus desired markup" the most popular pricing strategy, followed (nonexclusively) by watching other vendors and watching supermarket prices. Griffin and Frongillo (2003), in a series of interviews, found some farmers expect "coordination of prices" to be a standard—if informal—part of market life and express frustration at farmers who violate this code by competing on price. A survey of California vendors by Ahern and Wolf (2002) found vendors almost perfectly divided between those who vary their prices based on market conditions and those who do not.

One factor that may affect price is the number of competing vendors. The number of vendors in farmers markets does change somewhat during

β: LN(#VENDORS) VS. MEDIAN		β: LN(#VENDORS) VS. MINIMUM	
0.074	!!!	0.044	!!
0.096	!!!	-0.128	**
0.006	( )	-0.250	***
-0.245	***	-0.354	***
-0.194	( )	-0.201	( )
-0.277	***	-0.420	***
0.224	( )	-0.080	( )
-0.015	( )	-0.231	***
-0.059	***	-0.251	***
0.106	!!!	-0.010	( )

the year as vendors move in and out of the market during the growing seasons. One might expect that as the number of vendors competing in a farmers market increases, the price will fall, reflecting increased supply. Data collected for our case studies, which also included prices for leaf lettuce, were analyzed by individual farmers market by product combination. Twenty-three usable product by market combinations (e.g., "organic lettuce in Portland Market 1") were identified.

We regressed vendor numbers against both the median and minimum farmers market prices for a product on any given day. Median supermarket prices were included to control for seasonality and as a variable of interest in their own right. Twenty-one of the 23 markets analyzed had some statistically significant result using either the median or the minimum prices. Table 25 contains a selection of results, using the natural logarithms of all variables to allow insights about price elasticities.

Some market-product combinations, such as conventional spring mix in TwinCities1, behaved as conventional (economic) wisdom might predict: as more vendors entered the market, the median and minimum price levels declined. However, this result applies in just five market-product combinations. In another nine market-product combinations (e.g., Fuji apples in Portland1) only the minimum price declined. In other words, these results suggest that as the number of vendors in the market increases (new entries), one or more vendors offered lower prices, but these prices did not affect the general price level.

Six markets (e.g., ground beef in Syracuse) had statistically significant effects that contradicted the expected relationship: more vendors were associated with a higher median price. In two cases (certain apple varieties in Syracuse and Sacramento) the minimum price actually increased with vendor numbers. In these seemingly counterintuitive cases, a look at the original data often tells a similar story: at the peak of the season, some vendors raise prices or a new vendor enters with a significant markup—perhaps based on quality, to take advantage of large crowds, or because of a collusive dynamic.

Overall, two-thirds of markets did see a statistically significant decrease in the minimum price as vendor numbers increased. In other words, "undercutting vendors" do appear as markets grow, though consumers might have to shop around to find them.

Median prices decreased, however, in about one-third of markets. This result suggests that as the number of competitive vendors selling a particular product in any one market increases, the overall price level does not fall, as a general rule. This supports other survey findings that farmers often do not price on market supply but may use cost-plus markup techniques or set prices collectively. However, market and product attributes may play a role in the vendor-price effect: TwinCities1, for example, appears to be a relatively "competitive" farmers market; Syracuse does not. And competitive pricing is much more apparent in perishable products (lettuce and blueberries) than storable ones (frozen ground beef and apples). Thus vendors whose inventory can afford to "wait another day" are less likely to lower prices in the face of competition.

The data also allowed a statistical look at how farmers market and supermarket prices compare. In only one case (spring mix in Portland2) did the median farmers market price appear statistically "coupled" with the price in the conventional market. In some cases, they moved in opposite directions, and in most there was no relationship whatsoever. In addition, the results here do not demonstrate evidence of a general farmers market "discount" over supermarkets.

While in some ways these results affirm the notion of farmers markets as an enigmatic place to shop or do business, they begin the task of trying to discern some useful patterns amid the ever-busier stalls.

Other Product-Specific Price Model Insights

Table 26 presents results of the econometric model, which explains retail prices for the five focal products as a function of place of origin, season, retail outlet, price promotions, and product attributes. These results are employed to conduct a more nuanced analysis of the factors that influence prices of the focal products.

APPLES: Apple prices are somewhat seasonal, as summer prices are significantly greater than the prices during the rest of the year. Domestic stocks are generally low in the summer, and imports may be filling the demand gap. Regionally the Twin Cities has the highest apple prices. Price promotion and package size exhibit the expected effects: apples are sold at lower prices during promotions, and larger package sizes imply a lower per pound price for apples. Somewhat surprisingly, however, natural foods stores and farmers market apple prices are significantly lower than supermarket prices, even though the coefficient is relatively small. Bulk apples (apples sold individually rather than in a bag) are significantly more expensive than the rest. This is standard retail practice. Bulk apples are larger and of higher quality standard than bagged apples.

Apples are the only product of those studied that does not exhibit a price premium for local. Organic and select apple varieties do exhibit price premiums. A number of variety premiums are larger than the price premium for organic apples. The varieties Haralson, Pink Lady, and Honeycrisp all have premiums greater than the organic label.[4] Red Delicious is the base model variety.

BLUEBERRIES: In general blueberries are one of the highest priced products in the study. They are highly perishable, seasonal, and in increasing demand by consumers. The blueberry harvest season is mid to late

Table 26. Estimates from the econometric model

	APPLES ($ PER LB)	BLUE-BERRIES ($ PER LB)	SPRING MIX ($ PER LB)	GROUND BEEF ($ PER LB)	2% MILK ($ PER 1/2 GAL)
Mean Price	$1.66	$9.77	$8.99	$4.76	$3.18
N	3,732	1,105	2,014	2,073	1,607
R^2	0.762	0.546	0.765	0.713	0.790
Independent Variable	*Coefficient (Standard deviation)*				
Constant	1.220***	10.374***	13.220***	3.551***	2.067***
	(.042)	(.486)	(.129)	(.066)	(.041)
Fall	0.015	3.23***	-0.069	-.227***	-0.122***
	(.014)	(.355)	(.102)	(.051)	(.034)
Spring	-0.014	2.142***	0.087	-0.054	-0.033
	(.015)	(.362)	(.101)	(.053)	(0.034)
Summer	0.062***	-1.013***	0.086	-0.244***	-0.109***
	(.018)	(.394)	(.102)	(.052)	(.034)
Syracuse NY	-0.184***	-0.852**	0.344**	0.657***	0.246***
	(.036)	(.338)	(.139)	(.061)	(.036)
Washington DC	-0.132***	-0.613	-0.038	0.897***	0.129***
	(.035)	(.387)	(.118)	(.067)	(.045)
Portland OR	-0.415***	-1.799***	-1.154***	-0.157**	-0.035
	(.036)	(.323)	(.095)	(.061)	(.033)
Sacramento CA	-0.351***	1.04***	-0.706***	0.326***	0.357***
	(.036)	(.303)	(.092)	(.066)	(.042)
Price promotion	-0.597***	-3.992***	-1.857***	-0.678***	-0.363***
	(.022)	(.312)	(.138)	(.085)	(.034)

	APPLES ($ PER LB)	BLUE-BERRIES ($ PER LB)	SPRING MIX ($ PER LB)	GROUND BEEF ($ PER LB)	2% MILK ($ PER 1/2 GAL)
Independent Variable		*Coefficient (Standard deviation)*			
Package size	-0.047***	-0.247**	-0.474***	—	—
	(.005)	(.019)	(.008)		
Natural foods store	-0.092***	0.513**	0.417***	-0.005	0.016
	(.016)	(.254)	(.089)	(.049)	(.026)
Farmers market	-0.153***	0.336	-0.354	-1.120***	0.613***
	(.020)	(.414)	(.221)	(.080)	(.050)
Imported	-0.013	0.891***	—	—	—
	(.021)	(.299)			
Local	0.012	0.773**	2.254***	1.096***	0.334***
	(.016)	(.365)	(.195)	(.083)	(.032)
Organic	0.34***	2.477***	1.397***	2.254***	1.714***
	(.013)	(.231)	(.088)	(.063)	(.025)
Bulk-no packaging	0.668***	—	-0.136	—	—
	(.022)		(.117)		
% Lean (for Ground Beef)					
80% Lean	—	—	—	-0.801***	—
				(.060)	
90% Lean	—	—	—	0.709***	—
				(.058)	
95% Lean	—	—	—	0.924***	—
				(.072)	

	APPLES ($ PER LB)	BLUE-BERRIES ($ PER LB)	SPRING MIX ($ PER LB)	GROUND BEEF ($ PER LB)	2% MILK ($ PER 1/2 GAL)
Independent Variable			*Coefficient (Standard deviation)*		
Production Labels (for Ground Beef)					
Natural	—	—	—	0.582***	—
				(.054)	
Grass-fed	—	—	—	1.957***	—
				(.084)	
Varieties:					
Braeburn	-0.195***	—	—	—	—
	(.039)				
Empire	0.063**	—	—	—	—
	(.022)				
Haralson	0.380***	—	—	—	—
	(.054)				
Honeycrisp	1.200***	—	—	—	—
	(.035)				
Pink Lady	0.618***	—	—	—	—
	(.023)				
Gala/Fuji	0.324***	—	—	—	—
	(.038)				
Other	-0.018	—	—	—	—
	(0.134)				

*, **, *** = Significant at 10, 5, and 1 percent level, respectively.

summer, although the season varies slightly across the study regions. Prices are shown to vary significantly by season, with wider seasonal price swings than for the other four products. Portland and Syracuse, both in regions with strong local blueberry production, have significantly lower prices than the Twin Cities (the base for comparison) and the DC area, while Sacramento exhibits the highest blueberry prices. Portland has the advantage of being near significant commercial production and therefore has the lowest prices among the study regions.

SPRING MIX: Spring mix is a highly perishable product, and its observed retail prices by weight were also among the highest in this study. Prices do not vary seasonally, however, perhaps because the crops can be grown almost all year-round in a few major production areas of the United States. Retail prices do vary significantly across the study regions. Portland and Sacramento had significantly lower spring mix prices than the Twin Cities and the DC area; and Syracuse had higher prices. Price promotion and package size are negatively associated with retail prices. Considering price differences across retail outlets, natural foods stores sell spring mix at prices that are in general higher than prices in supermarkets and farmers markets. Prices of spring mix in farmers markets are not significantly different than prices in supermarkets. As mentioned earlier, the "local" and "organic" attributes receive price premiums relative to their standard counterpart. But, unlike for blueberries, the premium for "local" is much larger than the premium for "organic."

GROUND BEEF: Summer and fall prices of ground beef are significantly lower than in the winter and spring. Since 2009 was the middle of a depressed economy and generally saw extreme price competition in supermarkets, the study period may not reflect a typical seasonal or cyclical effect. The analysis also shows significant differences in prices across regions. Syracuse, the DC area, and Sacramento exhibit substantially higher prices for ground beef than the Twin Cities (the base for comparison); and Portland has the lowest prices among all study sites. Interestingly, ground beef prices in farmers markets are lowest among all retail outlets considered in the study. The model also suggests that the price of ground beef increases with the percent lean attribute.

Production labels are very important pricing factors, as both "natural" and "grass-fed" ground beef are more expensive than the "conventional" or unlabeled ground beef. Organically labeled ground beef has the highest price premium among all attributes, perhaps reflecting the higher cost of the systems producing and delivering organic meats. Local ground beef is significantly more expensive than its conventional counterpart; and the price premium of the attribute "local" falls between the premiums for "natural" and "grass-fed" production labels. If this is an indication of consumer preference, it appears that consumers are most interested in purchasing beef under specific, desirable production practices, with organic being the most preferred.

2 PERCENT MILK: Prices for 2 percent milk are affected by seasonality, although these differences are not as large as for other effects. In the study year of 2009, summer and fall retail prices were lower than winter and spring prices. However, the study year was a year when dairy farm prices plummeted sharply from historic highs to historic lows. Retail milk price patterns can vary significantly year to year, and this seasonal variation may be particular to this year only. Prices vary across regions; Syracuse, the DC area, and Sacramento have higher prices than the Twin Cities and Portland. Regarding retail outlets, prices of milk in farmers markets are significantly higher than prices in both supermarkets and natural foods stores. As expected, organic milk commands the highest price premium. Local milk is also more expensive but with a much smaller price premium than organic milk. We did not control for product attributes linked with specific production practice claims (such as natural, grass-fed, or hormone-free) in this analysis.

Conclusions

These analyses suggest that the local attribute provides a price premium for four of the five products: blueberries, spring mix, ground beef, and 2 percent milk. In addition, as expected, the organic attribute provides a price premium for all products. Other product attributes for specific products that carry price premium include variety for apples; and percent lean as well as grass-fed for ground beef.

Premiums calculated in this study are generally lower than those reported in willingness-to-pay studies. Consumers may overestimate their interest in local when presented with a survey as opposed to actual purchases. However, the results presented here should not be interpreted strictly without knowledge of the markets and of consumer behavior.

The premiums suggested in this study for the product attribute "local" hinge on the definitions of local used in these models, and changes in these definitions of local could alter the results. Definitions of local rely on consumer perceptions on what is local. In addition, consumers may have different perceptions as to what is local according to different products. For example, fluid milk is costly to transport long distances and would likely be labeled as local by many existing definitions, yet consumers do not think of milk purchased in the grocery store as a store brand as being a local product. And in general, milk packaging does not provide any information that would help to identify the milk as being locally produced or processed.

In some of the study regions (e.g., New York), some of the products are produced commercially for national and international markets (New York apples), yet are also viewed as local. It may be that consumers expect apples to be produced locally and therefore do not expect to pay a premium for the "local" attribute. Or perhaps, as pointed out with ground beef production labels, consumers are more interested in specific production methods and not in local production. The price premiums observed in these models using our definition of local may be linked more to perceptions of farm identity, farm size, production practices, label information, and marketing than to local geography.

Notes

1. This was confirmed in interviews with Oregon blueberry industry representatives.
2. These data are not available for Sacramento.
3. Data were collected for both whole and 2 percent milk.
4. These varieties may be perceived as "local" by consumers in specific markets, even when the apples do not satisfy the definition of "local" in those markets. For example, Haralson and Honeycrisp are often considered to be local apples by consumers in the Twin Cities.

References

Ahern, J., and M. M. Wolf. 2002. "California Farmers Markets Seller Price Perceptions: The Normative and the Positive." *Journal of Food Distribution Research* 33(1): 20–24.

Brown, C. 2003. "Consumers' Preferences for Locally Produced Food: A Study in Southeast Missouri." *American Journal of Alternative Agricultures* 18: 213–24.

Carpio, C. E., and O. Isengildina-Massa. 2009. "Consumer Willingness to Pay for Locally Grown Products: The Case of South Carolina." *Agribusiness* 25: 412–26.

Darby, K., M. T. Batte, S. Ernst, and B. E. Roe. 2008. "Decomposing Local: A Conjoint Analysis of Locally Produced Foods." *American Journal of Agricultural Economics* 90: 476–86.

Eastwood, D. B., J. R. Brooker, and R. H. Orr. 1987. "Consumer Preferences for Local Versus Out-of-State Grown Selected Fresh Produce: The Case of Knoxville, Tennessee." *Southern Journal of Agricultural Economics* 19 (December): 183–97.

Giraud, K. L., C. A. Bond, and J. J. Bond. 2005. "Consumer Preferences for Locally Made Specialty Food Products across Northern New England." *Agricultural and Resource Economics Review* 34: S75-S80.

Griffin, M. R., and E. A. Frongillo. 2003. "Experiences and Perspectives of Farmers from Upstate New York Farmers Markets." *Agriculture and Human Values* 20(2): 189–203.

Hinson, R. A., and M. N. Bruchhaus. 2005. "Louisiana Strawberries: Consumer Preferences and Retailer Advertising." *Journal of Food Distribution Research* 36: 86–90.

Logozar, B., and T. Schmit. 2009. *Assessing the Success of Farmers' Markets in Northern New York: A Survey of Vendors, Customers, and Market Managers*. Extension Bulletin, EB Series, *2009–2008* (55941). Department of Applied Economics and Management, Cornell University.

Loureiro, M. L., and S. Hine. 2002. "Discovering Niche Markets: A Comparison of Consumer Willingness to Pay for Local (Colorado Grown), Organic, and GMO-Free Products." *Journal of Agricultural and Applied Economics* 34: 477–87.

Offner, J. 2009. "Fresh Blueberry Volume Increasing in 2009." *The Packer*, May 5. thepacker.com/fruit-vegetable-news/marketing-profiles/berries/fresh_blueberry _volume_increasing_in_2009_122153809.html (accessed 1/26/2012).

USDA-AMS. 2010. "Farmers Market Growth, 1994–2010." August 4. U.S. Department of Agriculture, Agricultural Marketing Service. ams.usda.gov/AMSv1.0/farmersmarkets.

9 What Does Local Deliver?

Larry Lev, Michael S. Hand, and Gigi DiGiacomo

Throughout this volume we have presented insights from a set of case studies on the comparative performance of local and mainstream food production and distribution systems. This chapter poses the broad question: "What does local deliver?" and uses the case study findings and other available research to seek a response. Although the case studies do not yield observations along all possible dimensions of food system performance, they do suggest how supply chain structure and size are important determinants of local supply chain performance.

In the sections that follow, we examine how performance varies for supply chains that differ in terms of both geography and structure of operations. We begin with a "frustration ahead" warning. Although conflicting pronouncements touting either the benefits or the drawbacks of local are plentiful, the lack of a standard definition for local and the difficulty of tracking actual product movement make it impossible at this time to provide definitive answers to many key questions. Thus our response to "What does local deliver?" is frequently, "It depends." Still, the combination of the detailed supply chain case studies and other research does allow us to explore a set of key performance dimensions more precisely. The most fundamental conclusion presented in this chapter is that measuring the performance of supply chains—whether along environmental, social, economic, or other dimensions—is more about *how* food moves from producers to consumers than *where* (geographically) that food comes from. Even the comparison between supply chains that convey little information and those that convey detailed information to consumers about where, how, and by whom food was produced (a key component of the local supply chain definition used in this volume) does not yield consistent conclusions about most supply chain performance measures. A label of "locally produced" or a "local supply chain" gives no shortcut for predicting performance.

What Types of Performance Measures Are Important?

A lengthy list of impacts has been associated with increased localization of the food system.[1] These include revitalized local economies, reduced greenhouse gas emissions, preservation of farmland and rural lifestyles, decreased risks of food-borne illness or food-based threats to national security, and community building and democratization of the food system. Rather than address all potential performance impacts, we select a set of impacts that are central to the public debate related to local foods and that involve specific questions the supply chain case studies may help us answer. The topics in this essay cover a lot of ground but are clearly not a compendium of potential consequences of increased localization of the food supply. For example, the potential of local supply chains to affect social capital formation in a community or development of food democracy is left for others to consider.

Economic and Social Performance

This section examines the economic and social benefits and/or costs of local production, distribution, and consumption for communities and for producers. In discussing each, we begin with a brief review of applicable studies and then summarize the relevant case study findings.

Economic and Social Consequences for Local Communities

The growth in the number of farmers markets from 1,755 in 1994 to 6,132 in 2010 is a frequently cited indicator of both the increasing popularity of local foods and the value that individual communities place on participating in this movement.[2] Many consumers believe that purchasing locally produced agricultural products rather than products shipped in from elsewhere directly supports the local economy and local growers, and they indicate that these broader economic reasons are as important as the actual product characteristics in their decisions to buy local (Ono-zaka et al. 2010).

Tom Lyson (2004) coined the term *civic agriculture* to reflect the wide range of benefits he believed should be attributed to food systems that are tightly linked to the social development of communities: "Civic agriculture embodies a commitment to developing and strengthening an economically, environmentally, and socially sustainable agriculture and food production system that relies on local resources and serves local markets and consumers. The imperative to earn a profit is filtered through a set of cooperative and mutually supporting social relations. Community problem solving rather than individual competition is the foundation of civic agriculture" (Lyson 2004: 102).

Other observers, however, challenge these assessments and argue that the local food movement generates net negative consequences because "it is at odds with the principle of comparative advantage" (Lusk and Norwood 2011: 1).[3] Some suggest it can be dismissed as merely a way for "small subcategories of citizens . . . to find and express through the diets we adopt a solidarity with others who share our identities, our values, or our particular life circumstances" (Paarlberg 2010: 153–54). Unfortunately these critics of the local food movement focus on extreme scenarios (such as Lusk and Norwood's examination of the economic implications for North Dakota of producing everything, including pineapples, locally) and imply that government mandates and subsidies are driving the growth of local. In reality the dramatic growth of local food has been almost entirely demand-driven and therefore should be seen as a free market response to changing preferences within at least some consumer segments. Still, as indicated later in the chapter, the critics correctly identify important consequences of the localization of food systems that are almost completely ignored by other attempts to quantify the changes that occur.

Numerous studies examine the economic implications of the localization of food production and distribution within a specified area (e.g., Cantrell et al. 2006; Hughes et al. 2008; Kane et al. 2010; Swenson 2009; Tuck et al. 2010).[4] Each study identifies local economic benefits, ranging from the modest results reported for the actual impact of West Virginia farmers markets (Hughes et al. 2008: the 34 markets resulted in a net increase of $1.1 million in output and 43 full-time equivalent jobs) to much larger hypothetical impacts calculated for a six-state region (Swenson 2009: hundreds of millions of dollars in benefits and thousands of new jobs for dramatic increases in the seasonal production of fruits and vegetables in Illinois, Iowa, Michigan, Minnesota, and Wisconsin).

Three key areas for which assumptions vary across these studies are (1) whether the increase in local production for local markets results in a decrease in local production for nonlocal markets, (2) whether the increase in the consumption of local products results in a decrease in consumption of nonlocal products, and (3) whether the increased consumption of local products is market-driven, mandate-driven, or subsidy-driven. The assumptions made will directly establish the level of benefits calculated. For example, a study of the economic impact of farm-to-school programs in central Minnesota (Tuck et al. 2010) takes into account a wide range of income and employment impacts, including the reduced value of other crops in the local area as production acres are shifted, losses to distributors and retailers of nonlocal products no longer purchased, and the implications of higher prices paid for mandated local purchases. In this set of studies, including each element reduces but does not eliminate the economic benefits of going local.[5]

However, these studies fail to incorporate a final set of calculations required to analyze the implications of increased local sales at the national level. None of the individual "local" studies accounts for the losses suffered by supply chain participants outside the local area due to reduced sales. Ignoring these losses biases local economic impact studies. Consider the example of a country made up of two states, State A and State B, with all food production taking place in these two states and products moving freely between them. Assume that independently each state decides to study the economic implications of increased local sales. The State A model shows its own producers gaining sales at the expense of products formerly imported from State B, while the State B model shows a similar change taking place to the advantage of its locally oriented producers. Adding the results of the two models together would sum the local gains but ignore the nonlocal losses. The key implication is that other techniques such as the supply chain analyses used in the case studies in this volume are needed to capture a more complete set of consequences.

Several empirical studies have examined benefits that are not easily captured by the input-output studies. These include "clustering" benefits that result from sharing of information or infrastructure among locally oriented producers and spillover economic activity generated from farmers markets to neighboring businesses (see, for example, Gillespie et. al. 2007 and Lev et al. 2003). The valuable results of these focused studies provide an additional justification for combining multiple methods when assessing what local delivers.

Overall, our case studies collected only limited information related to the broad social and economic impacts that local supply chains have on communities. The direct market producers and the intermediated producers and supply chains did report stronger local community linkages than their mainstream counterparts. Larry Thompson set up farm stands on hospital campuses in metro Portland to contribute to the hospital goal of encouraging healthy eating. He also mentored a group of immigrant farmers. New Seasons Markets, a Portland area retailer, supports a wide range of local food system initiatives including farmers markets and farm stands that compete with it for consumer sales.[6] The Davis Food Co-op in California invests resources in strengthening local growers' entrepreneurial skills and incurs substantial transaction costs to purchase from local growers. In general, a higher proportion of wages and proprietor incomes is retained in the local area by the local food supply chains as compared to the mainstream supply chains.

Economic Implications for Producers

In recent decades there has been a strong trend toward fewer and larger farms providing an increasing percentage of total agricultural sales. Figure 30 shows that the share of sales contributed by farms with sales greater than $1,000,000, as depicted by the top box in each column, increased from 27 percent in 1982 to 59 percent in 2007. Would the expansion of local food systems slow or reverse this trend by providing additional support to the producer categories that have lost sales in recent decades?

Since no data set tracks product sales by the distance between production and consumption, it is not possible to use secondary data to determine whether and how changes in local sales might impact this trend or producer results more broadly. The Census of Agriculture does

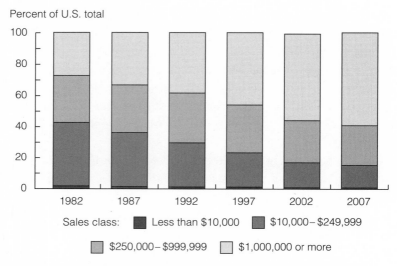

Percent of U.S. total

Sales class: ■ Less than $10,000 ■ $10,000–$249,999
■ $250,000–$999,999 ☐ $1,000,000 or more

Fig. 30. Market value of agricultural products sold by constant-dollar sales class, 1982–2007. Source: Hoppe et al. (2010).

provide data on farm-direct sales of products for human consumption, and a 2011 USDA study (Low and Vogel 2011) examines data for local intermediated sales. In the 2007 Census, 136,000 farms reported farm-direct sales of $1.2 billion, or an average of $8,800 in farm-direct sales per farm (USDA-NASS 2009). Most of these farms were quite small, as 80 percent had gross farm sales below $50,000 and nearly 95 percent had gross farm sales below $250,000. The farm-direct sector does, however, also include some larger farms. The nearly 7,500 farm-direct farms in the census with gross revenues above $250,000 averaged more than $68,000 in farm-direct sales. These larger farms contributed 43 percent of all farm-direct sales. In summary, these national data show that while most of the farms that sell through farm-direct channels are very small, a significant volume of farm-direct sales comes from commercial-sized operations.

Low and Vogel use the 2008 Agricultural Resource Management Survey to examine farm-direct and intermediated channel sales. They report that total local sales were $4.8 billion in 2008, with about 75 percent occurring in intermediated channels (Low and Vogel 2011: 111).

Table 27. Percentage difference in net producer price for local and mainstream supply chains

PRODUCT & LOCATION	% DIFFERENCE DIRECT MARKET VS. MAINSTREAM[a]	% DIFFERENCE INTERMEDIATED VS. MAINSTREAM[b]
Milk (DC Area)	+91%	+101%
Apples (Syracuse)	+50%	0%
Beef (Twin Cities)	+65%	+12%
Salad Mix (Sacramento)	+649%	+279%
Blueberries (Portland)	+183%	+194%

[a] Calculated as (Direct Market Net Price/Mainstream Net Price)−1.

[b] Calculated as (Intermediated Net Price/Mainstream Net Price)−1.

Source: Case studies.

Large farms, those with sales above $250,000, accounted for 92 percent of the sales that went exclusively through intermediated chains. These broader survey results mirror the case study results, which also show sales through intermediated channels dominated by farms that are quite a bit larger than the average farm-direct farm.

Aside from the case studies in this volume, there is little published research that compares the "net prices" (calculated as price received minus processing, distribution, and marketing costs) that growers receive in local and nonlocal supply chains.[7] Since producers in the local supply chains often assume responsibility for additional functions—including processing, distribution, and marketing, that can be costly and often require the operator's own labor—it is important to subtract these costs before comparing how well producers do when they sell through different supply chains.

Table 27 presents the percentage change in net prices received by case study producers in the direct and intermediated local supply chains as compared to the net prices received by mainstream producers. It shows that direct market producers earned greater (often much greater) net prices in all five case studies and that intermediated producers earned greater net prices in four of five instances. These higher prices are clearly a key attraction of the local market supply chains. The degree of variation in these comparative returns across the five cases is quite striking and indicates that great caution should be taken in trying to arrive at any rule of thumb for average differences between the net prices that producers receive in local and mainstream supply chains. It is also noteworthy that in two instances, milk in the DC area and blueberries in Portland, the intermediated producers' net prices slightly exceed those received by direct market producers.

Farm profitability is related to both net prices and volume.[8] Many local direct market producers operate in low-volume markets where price premiums can disappear if the delicate balance between supply and demand is upset. In some instances, a single new producer of a given product can cause this to happen.

So while these higher net prices were central to the success of the producers studied, they may not be available to new entrants or with further expansion by existing producers. In many instances, the intermediated markets have much greater opportunities for expansion (without major decreases in prices received by growers) than do the direct markets.

Environmental Performance

The current and potential contributions of local food systems to more environmentally sustainable agricultural systems continue to be discussed in academic and public forums. This discussion focuses primarily on whether local food supply chains reduce energy use and carbon emissions as compared to the mainstream food supply chain. The results are mixed. Some studies identify instances when local food products have the potential to reduce greenhouse gas emissions—for example, apples produced and consumed in Germany rather than imported from New Zealand (Blanke and Burdick 2005). Others have found that local products perform better for some products and not others—for example, carrot and tomato production in northern vs. southern Europe (Carlsson-Kanyama 1998). Given the existing research on the topic, the current assessment is that it is unclear whether local food supply chains perform better or worse on environmental measures.[9]

Food Miles Do Not Signal Environmental Performance

As defined in the case studies, local food supply chains are those that deliver products from nearby sources and convey information about where, how, and by whom a product was produced. Geographic proximity, the key information conveyed to consumers, is not an adequate proxy for better environmental performance. There is no particular reason to think that local supply chains perform better, or that encouraging further development of local food supply chains would result in improvements in overall environmental performance.

More specifically, "food miles" are not a reliable signal of the overall environmental performance of food supply chains or even of the performance of the transportation segments of supply chains (Pretty et al. 2005). Reducing food miles would reduce energy use, greenhouse gas emissions, and other environmental costs related to the transportation segments of supply chains if distance traveled were the only difference between mainstream supply chains and local supply chains. However, the supply chain case studies point out that transport modes, load sizes, relative fuel efficiency, and distribution logistics all differ among alternative supply chains and therefore also influence the relative environmental performance of supply chains. When differences in distances ("food miles") between mainstream supply chains and local supply chains are large, and differences in logistical efficiencies are small, local supply chains may reduce energy use and emissions.[10] But generalization beyond this special case is not possible.

Transportation segments in the case studies provide good examples of the complex relationships among marketing arrangements, geographic proximity, and environmental performance. The producers in the local supply chains that we studied recognized a market niche defined by direct connections with consumers. But several of the local supply chains had to forgo the transportation efficiencies gained in mainstream supply chains because the local chains operated at a smaller scale or could not maintain their identity in the marketplace if utilizing mainstream transportation networks. Some supply chains clearly choose to trade off more expensive (i.e., less energy efficient) transportation in order to maintain a marketing arrangement that they see as valuable.

In other cases, such as the Trickling Springs–MOM's Organic Market case in Washington DC, transportation efficiencies are related to both scale and the desire to convey information about a product's origin. While it is unlikely that Trickling Springs will be able to achieve the same level of transportation efficiencies that exist in mainstream milk supply chains, growth will likely provide incremental improvements in efficiency (and thus energy use) without sacrificing important product and supply chain characteristics.

A Broader Consideration of Environmental Performance

Even with the most optimistic assessment, the potential gains in environmental performance related to reducing transport distances may be small because transportation constitutes a relatively small portion of energy use and emissions in the U.S. food system. Transportation accounts for only about 3.5 percent of total energy use in food supply chains (Canning 2010). Furthermore total energy usage and emissions in transportation and other supply chain segments vary by product type.

The picture becomes more complicated when other segments of the supply chain are considered, because many factors in addition to geographic proximity determine performance. As an example, a study by Weber and Matthews (2008) found that overall red meat and dairy supply chains account for much larger total greenhouse gas emissions per household than do fruits and vegetables, even though produce transportation has higher per unit greenhouse gas emissions.

Production methods, size and economies of scale, transportation mode and load sizes, storage, processing, preparation, and waste can all differ between types of supply chains and can affect performance. Also, factors less related to supply chain characteristics can play a role: diet composition is a factor in environmental performance in the food system (Carlsson-Kanyama et al. 2003; Weber and Matthews 2008), and food waste represents a significant portion of energy consumption in the food system (Cuéllar and Weber 2010).

It is frequently asserted that food production and distribution are different in local food supply chains than in mainstream chains. Producers may use different production practices, may be more diversified, or may be a self-selected group of sustainability-minded farmers. Further, local supply chains may respond to consumers' desire for environmentally friendly production practices (i.e., a self-selected group of environmentally conscious consumers). Some of these differences in practices, particularly related to input use, can have an impact on environmental performance (see Lehuger et al. 2009; Kim and Dale 2008).

The supply chain cases note examples where production in a local supply chain is different from the corresponding mainstream chain and is potentially perceived as more environmentally friendly along certain dimensions: Shankstead's conversion to grass-based dairying, and grass-fed beef from Thousand Hills and the Brauchers.[11] But other local supply chains use production practices that are not distinguishable from mainstream supply chains. For example, South Mountain Creamery in DC and the local apple producers in New York use practices that are well within the norm for producers in mainstream supply chains yet are part of successful local supply chains that do not rely on differentiation by production practices. Therefore expansion of local foods will improve environmental performance only if the production practices that affect performance are integral to the local supply chain that develops.

Nutritional Health, Food Security, and Food Safety Performance

Nutritional health, food safety, and food security figure prominently in arguments for the increased localization of U.S. food systems. Consumer surveys suggest that buyers trust the term "local"—believing that food grown nearby is healthier, safer, and more secure. An Oregon Public Health Institute (2011) study highlights "access to healthy food" and "social connectivity" as two key social determinants of health that local food systems can help to provide. Martinez and colleagues conclude in their comprehensive review of literature, however, that "empirical evidence is insufficient to determine whether local food availability improves diet quality or food security" (Martinez et al. 2010: v). With this finding close at hand, we briefly discuss the potential health-related gains and losses offered by local foods and local food systems.

Nutritional Health

Pirog and Larson (2007: 23) report that 40 percent of U.S. consumers surveyed believe "Science has proven that local food is better for my health than food transported from across the country." While this perception that local food and local food supply chains provide fresher and more nutritious food when compared with mainstream alternatives is common (Dentoni et al. 2009; Darby et al. 2008; Onozaka et al. 2010), science has only begun to test these claims empirically. Ferrer and colleagues (2011), for example, applied regression techniques to explore the relationship between two diet-related diseases, diabetes and obesity, and local food intake among Georgia residents as measured by the number of farmers markets, farm direct sales, and vegetable production. Although this study found a significant but small inverse relationship between local food intake and diet-related disease, the direction of the causality remained unclear.

The nutritional health benefits of "going local" have also received theoretical attention. Edwards-Jones and colleagues (2008) state that local and nonlocal produce can be chemically analyzed by season for different supply chains and the resulting nutritional differences compared. Researchers hypothesize that differences in nutritional quality would depend on diverse factors including time of harvest, time from harvest to market, harvesting methods (machine v. hand), preservation technologies, handling methods, and shipping distance (Edward-Jones et al. 2008; Harvard Medical School Center for Health and Global Environment 2010). It is only when local food supply chains perform well across all of these factors that local products offer nutritional gains to the individual consumer. By itself and considered in isolation, the local attribute does not guarantee that nutritional benefits will be achieved.

This focus on the direct nutritional benefits associated with local food products is different from considering the broader question: "Can local food supply chains inspire positive, health-related, behavioral change in consumption decisions?" In other words, will consumers choose to eat healthier foods or maintain a healthier overall diet when access to locally produced foods or local food supply chains improves? This sort of behavioral change is a benefit frequently associated with farm-to-school programs and farmers markets. Farm-to-school programs, for example, typically include an educational component with the objective of creating long-term healthy eating behaviors. Farmers markets, too, have been a focus of public advocacy, with both policymakers and consumer groups arguing that increased access to these venues will improve fruit and vegetable intake and, in turn, inspire healthier eating habits.

North Carolina youth who live closer to farmers markets have been reported to have better body mass index scores than other youth (Jilcott et al. 2011). However, these correlation results should not be interpreted as proving that the existence of the markets leads to better nutritional outcomes. Instead, this may be a case of wealthier, more informed populations choosing to establish markets and/or farmers choosing to focus their sales in economically advantageous communities. Consumer surveys of nutrition assistance recipient populations suggest that fruit and vegetable intake increases with improved access and with nutrition assistance benefits (Anderson et al. 2001; Herman et al. 2008). But a detailed examination of these studies concludes that monetary incentives, in addition to access, appear to be necessary to generate increased fruit and vegetable purchases by the targeted population (McCormack et al. 2010). Similarly, we found no studies demonstrating that the presence of a farmers market leads to a net increase in the consumption of nutrient-dense foods or produce. Therefore we suggest that long-term improvements in diet may be more closely related to income, education, and social factors than to the local food supply chains or to the localness of the products themselves.

Food Security

Approximately 15 percent of American households are unable to acquire adequate food to meet their needs due to insufficient income or other resources (Nord et al. 2010). Food security is defined as "access at all times to enough food for an active, healthy life for all household members" (Nord et al. 2010: 2). Food access is largely a function of cost (as well as demographics, geography, and transportation infrastructure). As an example of how relative costs influence where consumers choose to shop, Ver Ploeg and colleagues (2009: iv) indicate: "Low- and middle-income households are more likely to purchase food at supercenters, where prices are lower."

Recent peer-reviewed price studies comparing food prices at supermarkets (mainstream supply chains) and farmers markets (local food supply chains) are limited.[12] Watson and Gunderson (2010) found that food for sale at Florida farmers markets was priced well below similar foods at nearby grocery stores. Similarly, Pirog and McCann (2009) compared local and nonlocal prices between farmers markets and supermarkets in the summer of 2009 using a market basket of foods commonly found at Iowa farmers markets. Findings from this study suggest some local price advantages, but the results overall are not statistically significant. Chapter 8 of this book presents an analysis of 51 weeks of 2009 price data collected at the five MSA locations in our case studies. Statistical results indicate that the "local" attribute increases prices for four of the five products studied (results for the fifth product, apples, are not statistically significant). Price premiums for the local products at all market locations range from 8.7 percent (blueberries) to 30.9 percent (ground beef). These results suggest that "local" may not be affordable—especially for the low-income households affected by food insecurity.[13]

With or without premiums, public assistance (e.g., coupons, discounts, education) appears necessary to stimulate local food purchases at farmers markets among low-income populations (Cole et al. 2007; McCormack et al. 2010). Coupon redemption programs, such as the Women, Infants, and Children Farmers Market Nutrition Program, generally increase fruit and vegetable intake among recipients and result in positive attitudinal changes toward produce consumption (McCormack et al. 2010). The research results suggest, however, that these behavior outcomes are short-lived and dependent upon ongoing public assistance.

The price effects of local food suppliers "scaling up" through increased production and centralized distribution are uncertain. At present, the price premiums charged for some local products conflict with the objective of improving food security for low-income populations. Thus we conclude that local food and local food supply chains do not appear to offer individual or household food security benefits beyond those of mainstream supply chains and may compromise the conditions of affordability.

Food Safety

National consumer surveys indicate that food safety is a significant concern among approximately half of the U.S. population (Deloitte 2011) and an important motivation for local food purchases (Onozaka et al. 2010; Pirog and Larson 2007). The Centers for Disease Control and Prevention (CDC) estimates that approximately one out of six Americans (48 million) become ill each year due to known food-borne pathogens and unspecified agents (CDC 2011). Illnesses occur when foods, usually raw, become contaminated by pathogens—parasites, bacteria, and viruses—that can enter the food supply at any point in the supply chain (Council to Improve Foodborne Outbreak Response 2009).

Will consuming a higher percentage of local foods reduce public exposure to food-borne pathogens? We suggest that local food and local food supply chains theoretically perform no better (or worse) than nonlocal food distributed through mainstream supply chains when safe production, handling, and manufacturing practices are followed. In other words, food produced and distributed locally is as susceptible to contamination as food grown and distributed through mainstream supply chains. Once an illness or outbreak occurs, however, local and mainstream supply chains differ in their ability to manage and respond to the threat. Local food supply chains often provide consumers with greater transparency and thus traceability by conveying information about where, how, and by whom the product was produced.

Moreover, smaller, intermediated supply chains and direct market supply chains inherently limit the scope of food-borne pathogen outbreaks due to limited market penetration and their limited scale. The production, processing, and distribution efficiencies associated with mainstream supply chains that make it possible for large volumes of food to move long distances also mean that contaminated food can be widely distributed before contamination has been detected. The benefits associated with local food supply chains, however, would be compromised should the scope of distribution grow through aggregated supply and distribution. In other words, all other things being equal, scaling up to achieve greater market penetration would reduce any inherent limit on the impact of outbreaks. Thus when asking whether local food supply will improve food safety, we return to our earlier statement—it depends, in this instance on the scale and degree of aggregation achieved.

Concluding Comments

Any consideration of what local delivers must begin with the recognition that it provides satisfaction to a certain segment of consumers and citizens. Beyond that, the deliverables are murkier. The social benefits to communities are less controversial than the economic and environmental benefits. Locally oriented, generally smaller producers realize some economic advantages from the expansion of local food systems.

The case studies make clear (as have other researchers) that distance alone is insufficient to assess the environmental performance of local and mainstream food distribution systems. Other factors related to the production and distribution systems must be considered.

The review of research with respect to nutritional and health outcomes yields a similar conclusion. The localness of products is a weak determinant of the nutritional composition, adequacy, and safety of consumer food purchases.

These findings suggest that much is still unknown about "what local delivers." As local food becomes more prevalent in the U.S. food system, additional research can provide a clearer picture of the benefits and costs of emerging local food supply chains. Of particular interest is research that can better quantify the net impacts of expanding local food markets at regional, national, and international scales. A related area for future research might compare how supply chains perform in rural and urban areas. The case studies examined supply chains with significant connections to both urban and rural areas; it is not clear whether local food supply chains perform differently when they are primarily based in and serve rural communities.

Further, little is known about how producers, processors, and other enterprises fare when participating in local food supply chains. The case studies provide some evidence of how producer returns vary by supply chain type, but more systematic analyses have not yet been undertaken. Although there is a long history in agricultural economics of research on producer costs, returns, and farm structure, it may be useful to examine how different supply chain arrangements may relate to these outcomes.

Research in other areas discussed in this essay—environmental, health and nutrition, and food safety outcomes—may benefit from a shift in perspective toward how food reaches consumers, rather than its geographic origin. How food is processed and stored, transported, and marketed all have important roles to play in determining the environmental impact, quality, quantity, and price of foods available in any community. These aspects of what local delivers, along with economic impacts, broadly describe the various facets of sustainability in the food system. Research on the role of supply chain arrangements in affecting measures of sustainability can provide insight on performance tradeoffs that may exist as local food supply chains make up a larger portion of the food sector.

The existing research findings that we highlight make clear that "local" does not unambiguously translate into "more sustainable" along several measures. We suspect that this is because the many ways that food can move from farms to consumers, in both local and mainstream supply chains, can confound the true determinants of what food supply chains deliver. Untangling supply chain arrangements from geographic distance and origin improves our understanding of how performance on sustainability measures varies in the food system.

Notes

1. Throughout the discussion, we focus on relatively small increases in local food production and distribution rather than examining the impacts of completely replacing the mainstream food distribution system with local food.

2. Stephenson et al. (2008) report, however, that for the period 1998 to 2005 nearly 50 percent of new Oregon markets failed within the first four years of operation. So the interest in local does not mean that all local food ventures will succeed.

3. Under comparative advantage, regions or countries gain by specializing in producing what they do relatively more efficiently and trading for other products.

4. These studies use input/output models to examine the direct, indirect, and induced benefits of either hypothetical or actual changes in production and consumption.

5. Other scenarios could certainly result in net losses for going local.

6. New Seasons Market: newseasonsmarket.com/our-community/welcome-to-our -community (accessed 2/13/2012).

7. Most other case studies simply report the actual prices received by growers from different supply chains and do not take the next step of subtracting out the marketing costs associated with sales through each supply chain.

8. Production costs also influence profitability. We did not, however, calculate production costs in these case studies.

9. See Martinez et al. (2010: 48–49) for a review of relevant studies.

10. See, for example, Pirog et al. 2001.

11. We did not specifically evaluate environmental performance of production systems. However, some life-cycle analyses indicate that grass-fed production systems do not necessarily yield better outcomes for greenhouse gas emissions and other

environmental performance measures. See, for example, Pelletier et al. (2010) and Peters et al. (2010).

12. Earlier studies from the 1970s–1980s observed lower prices at farmers markets when compared with nearby supermarkets. However, "Few writers analyze data according to the class of market being discussed; wholesale and retail farmers markets sometimes mingled in the same data sets" (Brown 2002: 170).

13. Chapter 10, "Can Local Food Markets Expand," notes that these premiums are below those reported in willingness-to-pay studies.

References

Anderson, J., D. Bybee, R. Brown, D. McLean, E. Garci, M. Breer, and B. Schillo. 2001. "Five a Day Fruit and Vegetable Intervention Improves Consumption in a Low Income Population." *Journal of American Dietetic Association* 101: 195–202.

Blanke, M. M., and B. Burdick. 2005. "Food (miles) for Thought." *Environmental Science and Pollution Research* 12(3): 125–27.

Brown, A. 2002. "Farmers' Market Research 1940–2000: An Inventory and Review." *American Journal of Alternative Agriculture* 17(4): 167–76.

Canning, P. 2010. "Fuel for Food: Energy Use in the U.S. Food System." *Amber Waves* 8(3): 1–6.

Cantrell P., D. Conner, G. Erickcek, and M. Hamm. 2006. *Eat Fresh and Grow Jobs, Michigan.* Beulah: Michigan Land Use Institute. www.mlui.org/userfiles/filemanager/274/.

Carlsson-Kanyama, A. 1998. "Food Consumption Patterns and Their Influence on Climate Change: Greenhouse Gas Emissions in the Life-Cycle of Tomatoes and Carrots Consumed in Sweden." *Ambio* 27(7): 528–34.

Carlsson-Kanyama, A., M. P. Ekstrom, and H. Shanahan. 2003. "Food and Life Cycle Energy Inputs: Consequences of Diet and Ways to Increase Efficiency." *Ecological Economics* 44: 293–307.

Centers for Disease Control. 2011. CDC *Estimates of Foodborne Illness in the United States.* CS218786-A. February.

Cole, B., S. Hoffman, R. Shimkhada, and C. Rutt. 2007. "Health Impact Assessment of Modifications to the Trenton Farmers Market." UCLA Health Impact Assessment Group. March 20.

Coley, D., M. Howard, and M. Winter. 2009. "Local Food, Food Miles and Carbon Emissions: A Comparison of Farm Shop and Mass Distribution Approaches." *Food Policy* 34: 150–55.

Council to Improve Foodborne Outbreak Response (CIFOR). 2009. *Guidelines for Foodborne Disease Outbreak Response.* Atlanta: Council of State and Territorial Epidemiologists.

Cuéllar, A. D., and M. E. Weber. 2010. "Wasted Food, Wasted Energy: The Embedded Energy in Food Waste in the United States." *Environmental Science and Technology* 44(16): 6464–69.

Darby, K., M. T. Batte, S. Ernst, and B. E. Roe. 2008. "Decomposing Local: A Conjoint Analysis of Locally Produced Foods." *American Journal of Agricultural Economics*. 90: 476–86.

Deloitte. 2011. *Consumer Food Safety Survey Results*. New York: Deloitte Development, LLC.

Dentoni, D., G. Tonsor, R. Calatone, and H. Peterson. 2009. "The Direct and Indirect Effects of 'Locally Grown' on Consumers' Attitudes towards Agri-Food Products." *Agricultural and Resource Economics Review* 38(3): 384–96.

Edwards-Jones, G., L. M. I. Canals, N. Hounsome, M. Truninger, G. Koerber, B.Hounsome, P. Cross, E. H. York, A. Hospido, K. Plassmann, I. M. Harris, R. T. Edwards, G.A.S. Day, A. D. Tomos, S. J. Cowell, and D. L. Jones. 2008. "Testing the Assertion that 'Local Food Is Best': The Challenges of an Evidence-Based Approach." *Trends in Food Science and Technology* 19: 265–74.

Ferrer, M., E. Fonsah, O. Ramirez, and C. Escalante. 2011. "Local Food Impacts on Health and Nutrition." Paper presented at Agricultural and Applied Economic Association 2011 AAEA and NAREA Joint Annual Meeting, July 24–26, Pittsburgh .

Food and Drug Administration. 2010. *FDA Consumer Health Information*. December.

Food and Drug Administration. 2010. *Retail Food Risk Factor Study*. October.

Gerald, B. L. 2009. "Safety Issues in the U.S. Food System." In *Food and Nutrition at Risk in America: Food Insecurity, Biotechnology, Food Safety, and Bioterrorism*, ed. S. Edelstein, B. L. Gerald, T. C. Bushnell, and C. Gundersen, 41–56. Sudbury MA: Jones and Bartlett Publishers.

Gillespie, G. Jr., D. L. Hilchey, C. C. Hinrichs, and G. Feenstra. 2007. "Farmers' Markets as Keystones in Rebuilding Local and Regional Food Systems." In *Remaking the North American Food System*, ed. C. Clare Hinrichs and Thomas A. Lyson, 65–83. Lincoln: University of Nebraska Press.

Harvard Medical School Center for Health and Global Environment. 2010. "Is Local More Nutritious? It Depends." Center for Health and Global Environment, Harvard Medical School. chge.med.harvard.edu/programs/food/nutrition/html (accessed 4/13/2011).

Herman, D., G. Harrison, A. Afifi, and E. Jenks. 2008. "Effect of a Targeted Subsidy on Intake of Fruits and Vegetables among Low-Income Women and Children in the Special Supplemental Nutrition Program for Women, Infants and Children." *American Journal of Public Health* 98: 98–105.

Hoppe, R. A., J. M. MacDonald, and P. Korb. 2010. *Small Farms in the United States: Persistence under Pressure*. EIB 63. Washington DC: U.S. Department of Agriculture, Economic Research Service.

Hughes, D.W., C. Brown, S. Miller, and T. McConnell. 2008. "Evaluating the Economic Impact of Farmers' Markets Using an Opportunity Cost Framework." *Journal of Agricultural and Applied Economics* 40: 253–65.

Jilcott, S., S. Wade, J. McGuir, Q. Wu, S. Lazorick, and J. Moore. 2011. *The Association between Food Environment and Weight Status among Easter North Carolina Youth*. Public Health Nutrition. February. doi:10.1017/s1368980011000668.

Kane, S. P., K. Wolfe, M. Jones, and J. McKissick. 2010. "The Local Food Impact: What if Georgians Ate Georgia Produce?" Report CR-10-03. Center for Agribusiness and Economic Development, College of Agricultural and Environmental Sciences, University of Georgia. caes.uga.edu/unit/oes/documents/LocalFoodImpact_july2_2010.pdf.

Kim, S., and B. E. Dale. 2008. "Effects of Nitrogen Fertilizer Application on Greenhouse Gas Emissions and Economics of Corn Production." *Environmental Science and Technology* 42(16): 6028–33.

Lehuger, S., B. Gabrielle, and N. Gagnaire. 2009. "Environmental Impact of the Substitution of Imported Soybean Meal with Locally-Produced Rapeseed Meal in Dairy Cow Feed." *Journal of Cleaner Production* 17: 616–24.

Lev, L., L. Brewer, and G. Stephenson. 2003. *How Do Farmers' Markets Affect Neighboring Businesses?* Oregon Small Farms Technical Report No.16, Oregon State University Extension Service. smallfarms.oregonstate.edu/sites/default/files/publications/techreports/TechReport16.pdf.

Levi, L., S. Vinter, R. St. Laurent, and L. Segal. 2010. *F as in Fat: How Obesity Threatens America's Future*. N.p.: Trust for America's Health and Robert Wood Johnson Foundation.

Low, Sarah, and Stephen Vogel. 2011. *Direct and Intermediated Marketing of Local Foods in the United States*. ERR-128. Washington DC: U.S. Department of Agriculture, Economic Research Service.

Lusk, J. L., and F. B. Norwood. 2011. "The Locavore's Dilemma: Why Pineapples Shouldn't be Grown in North Dakota." *Library of Economics and Liberty*. econlib.org/library/Columns/y2011/LuskNorwoodlocavore.html.

Lyson, T. 2004. *Civic Agriculture: Reconnecting Farm, Food, and Community*. Boston MA: Tufts Press.

Martinez, S., M. Hand, M. Da Pra, S. Pollack, K. Ralston, T. Smith, S. Vogel, S. Clark, L. Lohr, S. Low, and C. Newman. 2010. *Local Food Systems: Concepts, Impacts, and Issues*. ERR-97. Washington DC: U.S. Department of Agriculture, Economic Research Service.

McCormack, L. A., M, N. Laska, N. Larson, and M. Story. 2010. "Review of the Nutritional Implications of Farmers' Markets and Community Gardens: A Call for Evaluation and Research Efforts." *Journal of the American Dietetic Association* 110(3): 399–408.

Mukiibi, M., J. Bukenya, J. Molnar, A. Siaway, and L. Rigdon. 2006. "Consumer Preferences and Attitudes towards Public Markets in Birmingham, Alabama." Paper presented at Southern Agricultural Economics Association Annual Meeting, February 5–8, Orlando, Florida.

Nord, M., A. Coleman-Jensen, M. Andrews, and S. Carlson. 2010. *Household Food Security in the United States, 2009*. ERR-108. Washington DC: U.S. Department of Agriculture, Economic Research Service.

Onozaka Y., G. Nurse, and D. T. McFadden. 2010. "Local Food Consumers: How Motivations and Perceptions Translate to Buying Behavior." *Choices* 25(1).

Oregon Public Health Institute. 2011. "Healthy Eating at Farmers Markets: Findings from Reflection and Listening Sessions." http://www.ophi.org/download/PDF/healthy _planning_pdfs/HEFMtour_web.pdf.

Paarlberg, R. 2010. *Food Politics: What Everyone Needs to Know.* New York: Oxford University Press.

Pelletier, N., R. Pirog, and R. Rasmussen. 2010. "Comparative Life Cycle Environmental Impacts of Three Beef Production Strategies in the Upper Midwestern United States." *Agricultural Systems* 103: 380–89.

Peters, G. M., H. V. Rowley, S. Wiedermann, R. Tucker, M. D. Short, and M. Schultz. 2010. "Red Meat Production in Australia: Life Cycle Assessment and Comparison with Overseas Studies." *Environmental Science and Technology* 44: 1327–32.

Pirog, R., and A. Larson. 2007. *Consumer Perceptions of the Safety, Health, and Environmental Impact of Various Scales and Geographic Origin of Food Supply Chains.* Ames IA: Leopold Center for Sustainable Agriculture.

Pirog, R. and N. McCann. 2009. *Is Local Food More Expensive? A Consumer Price Perspective on Local and Non-Local Foods Purchased in Iowa.* Ames IA: Leopold Center for Sustainable Agriculture.

Pirog, R., T. Van Pelt, K. Enshayan, E. Cook. 2001. *Food, Fuel, and Freeways: An Iowa Perspective on How Far Food Travels, Fuel Usage, and Greenhouse Gas Emissions.* Ames IA: Leopold Center for Sustainable Agriculture.

Pretty, J. N., A. S. Ball, T. Lang, and J.I.L. Morison. 2005. "Farm Costs and Food Miles: An Assessment of the Full Cost of the UK Weekly Food Basket." *Food Policy* 30: 1–19.

Stephenson, G., L. Lev, and L. J. Brewer. 2008. "Things Are Getting Desperate: What We Know about Farmers' Markets that Fail." *Renewable Agriculture and Food Systems* 23: 188–99.

Swenson, D. 2009. *Investigating the Potential Economic Impacts of Local Foods for Southeast Iowa.* Ames IA: Leopold Center for Sustainable Agriculture.

Toler, S., B. Briggeman, J. L. Lusk, and D. C. Adams. 2009. "Fairness, Farmers Markets, and Local Production." *American Journal of Agricultural Economics* 91(5): 1272–78.

Tuck, Brigid, Monica Haynes, Robert King, and Ryan Pesch. 2010. "The Economic Impact of Farm-to-School Lunch Programs: A Central Minnesota Example." University of Minnesota Extension.

USDA-NASS. 2009. *Census of Agriculture, 2007.* Washington DC: U.S. Department of Agriculture, National Agricultural Statistics Service.

Ver Ploeg, M., V. Breneman, T. Farrigan, K. Hamrick, D. Hopkins, P. Kaufman, B. H. Lin, M. Nord, T. Smith, R. Williams, K. Kinnison, C. Olander, A. Singh, and E. Tuckermanty. 2009. *Access to Affordable and Nutritious Food: Measuring and Understanding Food Deserts and Their Consequences.* Report to Congress, June. Washington DC: U.S. Department of Agriculture, Economic Research Service

Watson, J., and M. Gunderson. 2010. "Direct Marketing of Specialty Crops by Producers: A Price Comparison between Farmers' Markets and Grocery Stores." Paper presented at Southern Agricultural Economics Association Annual Meeting, February 5–8, Orlando, Florida.

Weber, C. L., and H. S. Matthews. 2008. "Food-Miles and the Relative Climate Impacts of Food Choices in the United States." *Environmental Science and Technology* 42(10): 3508–13.

10 Can Local Food Markets Expand?

Edward W. McLaughlin, Shermain D. Hardesty, and Miguel I. Gómez

Chapter 9 discussed the challenges of assessing the performance of local food supply chains relative to their mainstream counterparts. In this essay we turn our attention to the question: "Can local food markets expand?" Consumers' growing interest in foods with a variety of attributes associated with being "locally produced" provides the rationale for raising this question. We argue that a very large expansion of sales through direct markets is likely to be infeasible from a consumer perspective. In fact, a recent USDA study estimated that of the $4.8 billion in sales of locally grown foods marketed in the United States during 2008, the value of intermediated sales was more than three times higher than the value of local foods marketed exclusively through direct-to-consumer channels (Low and Vogel 2011). Therefore the expansion of local foods is likely to depend, to a great extent, on the ability of local food supply chains to create links with their mainstream counterparts. We examine two interrelated questions throughout the supply chain that need to be considered when assessing the expansion potential of local foods: (1) Would local supply chains be able to facilitate the flow of information of the attribute "local" and implement the appropriate pricing strategies in order to achieve the minimum scale required to enable local foods to compete with mainstream foods? And (2) Are the obstacles associated with food safety regulation, seasonality in production, and producers' lack of entrepreneurial skills surmountable? In this essay we describe how each of these factors impacts the growth potential for local foods, and we offer suggestions on how to overcome the related problems emphasizing the advantages of mainstream supermarkets and natural food cooperatives to increase the share of local foods in total food consumption relative to direct market supply chains.

Information Flow Is Critical

Food supply chains are facing new demands from consumers and regulators to provide product-specific characteristics beyond simply price, availability, and nutrient composition. Increased consumer interest in local foods is one of the most salient of such new demands. To address this increased interest, food firms are attempting to develop new strategies and distribution channels to benefit from the value consumers attach to the attribute "local." But contrary to other attributes, which are often easier to define (e.g., price and size) and certify (e.g., organic), the "local" attribute has different meanings for different consumers and supply chain participants. "Local" is often conflated with other attributes, such as those associated with small family farms, healthier products, and sustainable natural resource and energy use, among others. The 2007 Farm Bill defines *local* as "the locality or region in which the final product is marketed, so that the total distance that the product is transported is less than 400 miles from the origin of the product; or the State in which the product is produced."[1] This diversity of definitions generates a wealth of opportunities for product differentiation among food producers and marketers. But the same diversity could confuse consumers and ultimately erode the value that they attach today to the attribute "local." Some fear that food firms may even abuse this powerful but elusive attribute through the use of dubious marketing claims.

The lack of clarity about what local means highlights the importance of credible and meaningful information flow regarding local foods among supply chain members, including consumers. In our case studies we find that direct market supply chains generally offer consumers detailed information about where, by whom, and how the product is produced. However, growing the share of local foods depends on the ability of the food system to link growers of local products to mainstream food supply chains. One particular challenge is how to convey appropriate information about local products effectively as the supply chain adds local intermediaries. For example, Market of Choice, a retailer in Eugene, Oregon, uses innovative ways to facilitate the flow of information about the local foods it sells; in addition to having a well-structured program for local sourcing, this retailer often displays videos of local growers and features them in each store department (Wright 2010).

Information flow about local foods may be easier to convey for processed foods, through use of various types of packaging, but is more difficult for unpackaged fresh foods like most produce or meats. Nevertheless, a growing number of retailers require their suppliers (packers, shippers, and distributors) to provide grower identities as part of the criteria for retailers' food safety traceability programs. It is therefore possible that in certain cases, this information can be used to incorporate the attribute "local" in the supermarket's marketing communication strategy. This would require additional efforts to improve packaging as well as coordination between local growers and mainstream retailers. Specifically, retailers could employ more specific information developed in cooperation with local growers and could conduct more activities to communicate this information actively with greater authenticity to their shoppers.

Customer education is one of the distinguishing features of natural foods cooperatives; one of the seven principles adopted by most of these cooperatives is to provide education to their members so that they can contribute effectively to the development of their cooperatives. The board policies of the Davis Food Co-op, the focus of the intermediated spring mix case, include: "Advance an educational shopping environment . . . extensive product signage and labeling that highlight product sources, systems of production and harvest. . . . Offer newsletters, web pages, classes, speakers, events, books and advertising that include educational components" (Davis Food Cooperative 2011). This has included displaying its local food logo for products grown within 100 miles of Davis and often identifying the specific farm for local produce items.

Firms in mainstream supply chains could invest in electronic information technologies to allow consumers to use in-store scanners to review grower information during shopping trips; indeed, some are already well advanced in this area. In Japan, quick response codes (QR codes) are displayed on packaged fruits and vegetables; consumers can scan the codes with their smartphones and learn by whom, where, and how the produce was grown (Mendell 2009). Similarly, grocers could display QR codes for locally grown produce that is sold unpackaged. Growers, for their part, could embrace new social media tools to convey information directly to target shoppers and to promote their products at grocery stores. The challenge to industry participants is to devise mechanisms to facilitate new information flow in a cost-effective manner. This is an important consideration because information is often costly to convey. Therefore the quantity demanded as well as the willingness to pay for local foods need to be sufficient to justify investments in information conveyance. Already some supermarkets are employing Twitter to send messages to customers letting them know the exact time that local sweet corn is delivered to the store and put on display. Growers could build similar databases from their customer loyalty programs and could follow similar practices.

Aggregation

As noted by King and colleagues (2010), significant growth in local foods sales requires that producers supply an adequate volume of competitively priced products. This typically requires aggregation of product across producers. Aggregation of local foods during processing, distribution, and marketing activities is critical for achieving scale economies and reducing transaction costs of system participants, especially the consumers. A farmers market can be considered the most simplistic form of aggregation, since it is a venue that brings together producers and consumers to transact sales. Newly formed entrepreneurial food companies can also involve product aggregation. Thousand Hills Cattle Company, the intermediated beef case discussed in chapter 6, raises some of its cattle, but it also purchases finished animals that conform to Thousand Hills' strict production protocol from approximately 40 producers.

Historically, farmers have formed cooperatives to aggregate production and benefit from economies of scale in processing, transportation, and marketing. The Maryland and Virginia Milk Producers Cooperative Association, described in chapter 7 in the mainstream fluid milk case, is such an organization. By joining a processing cooperative, producers no longer need to search for a processing facility, nor do they need to negotiate the processing charges individually. Moreover, they do not have to bear the entire burden of the investment costs and risks. Distributors and retailers who purchase from the cooperative have to submit only one order, process only one invoice, and handle only one delivery from the cooperative, rather than having to manage individual orders, invoices, and deliveries from each of the producer-members who supplied the purchased products. Similarly, consumers form natural foods cooperatives, such as the Davis Food Cooperative in the intermediated spring mix case described in chapter 5, to be able to purchase foods with key desired characteristics at a price lower than they would pay if they ordered the items as individuals. Furthermore, some natural foods cooperatives have enhanced their aggregation capabilities by establishing distribution warehouses, such as The Wedge's Co-op Partners Warehouse located in St. Paul. Thousand Hills Cattle Company, the intermediated beef case discussed in chapter 6, makes deliveries to Co-op Partners Warehouse, where the beef is aggregated with other products and transported to natural foods cooperatives in the area.

The aggregation activities performed by Co-op Partners Warehouse and the Maryland and Virginia Milk Producers Cooperative Association are both examples of the so-called food hubs, or "centrally located facilities with a business management structure facilitating the aggregation, storage, processing, distribution, and/or marketing of locally/regionally produced food products" (USDA 2010). Small- and medium-scale farmers can therefore gain access to capital- and knowledge-intensive marketing services such as storage, processing, and marketing services, among others.

Aggregation can also be achieved through other organizational structures. Red Tomato is a nonprofit produce marketing organization that generates economies of scale in transportation and marketing and reduces producers' and customers' transaction costs. In its early days Red Tomato owned a warehouse and operated trucks to aggregate and distribute farmers' products in New England, New York, and the mid-Atlantic. After several years of operation, it determined that this structure was not financially viable and disposed of these assets. Today Red Tomato's objective is to coordinate, rather than operate, the supply chain by utilizing its expertise in marketing and logistics. It now consolidates product at centrally located farms, at produce markets, and at distribution centers. Deliveries are made either directly to stores or distribution centers by several farmers and trucking companies (Stevenson 2009). Farmers markets have been recognized as serving a critical role in aggregating product for consumers. However, the popular Santa Monica Farmers Market in Southern California, which hosts 75 to 80 farmers every Wednesday, also aggregates produce for commercial customers. Buyers from local produce distributors shop for five to 25 customers each week and have been sourcing from the market for five to 15 years. They purchase between $1,000 and $4,000 of produce each week from the farmers market. Additionally, one grower reported that about 70 percent of the product that he brings to the farmers market has been presold to restaurants (Dawson 2012).

Alchemist is a community development corporation in Sacramento, California, that purchases produce from several local farms to be marketed at farm stands it operates in low-income "food deserts" within the city. In other cases, farmers are developing their own aggregation firms. For example, Capay Valley Growers is a partnership involving approximately 25 local farms and ranches that was formed to aggregate their produce and processed food products and make weekly deliveries to subscribers at worksites and other locations, similar to a community supported agriculture program.

Such aggregation makes it more difficult and more costly to retain product information, which as previously discussed is important to consumers of local foods. Although federal and state pack-and-grade standards for produce sold through wholesale channels already require that each grower or packer-shipper packs its products in boxes marked with the brand and location (city and state), this information is not usually provided to consumers at mainstream grocery stores. The primary function of packer-shippers is to aggregate deliveries of a specific fruit or vegetable, such as peaches or lettuce, from several growers. As noted in the preceding section, technology now exists to simplify the flow of information to consumers regarding where, how, and by whom a product was produced. Each box of produce packed by many packer-shippers is stamped with a code or a label that can be used to trace the product back to a specific grower.

The produce category provides an interesting illustration of such information flow. Much of the produce sold by Nugget Markets, the mainstream retailer in the spring mix case, is supplied by its produce distributor, which buys from numerous packer-shippers. However, Nugget has begun to feature local farmers. The growers typically deliver their produce directly to Nugget's produce distributor, rather than marketing through a packer-shipper. For example, Vierra Farms grows organic watermelons, heirloom tomatoes, and squash on a local 100-acre parcel. David Vierra approached Nugget about marketing his watermelon and volunteered to talk with customers at select Nugget stores. He recently stood out in front of a Nugget store offering samples of his black and yellow watermelons, helping customers pick out melons, and sharing the story of his family's fifth generation farm. Nugget's produce manager was pleased with the response and is seeking out other growers who will interact with customers at two or three of their stores where interest in buying locally grown foods is greatest.

It should be noted that such aggregation efforts are not brought about easily. Cooperatives have long offered certain economic advantages, but agricultural producers have not always been willing to cede their independence and their own perceived brand identity to become part of a group. Moreover, the management skills required to organize fragmented producers are not currently in evidence. Similarly, the very act of aggregation into larger volumes risks diluting the appeal of the local attribute for the retailers seeking such competitive advantage and for the very consumers who enjoy relationships with individual small farmers.

Pricing Strategies for Local Foods Are Challenging

Initiatives to facilitate the flow of information and increase demand for local foods raise questions about the costs that such efforts may add to the supply chain. Would prices received by local growers from supermarkets compensate for the higher costs of conveying additional information? Pricing is often the most difficult variable to manage in the marketing mix for any product or service, and local foods are no exception. Many local growers have relied extensively on direct marketing to sell their products at full retail prices and to capture a larger share of the retail value. Our case studies indicate that growers are able to earn higher returns in direct market chains, even after adjusting for the often considerably higher marketing costs they incur, relative to other supply chains. It is generally believed that local growers require higher shares of retail value in order to cover the greater production costs stemming from their highly diverse crop mixes on relatively small acreages. But focusing solely on marketing margins may be a mistake.

We believe that increasing the share of local products requires a fundamental change in the business models adopted by many local growers. In particular, local growers may be able to increase sales volumes by accepting reduced marketing margins. Our case studies illustrate that local growers sometimes refuse to accept standard wholesale prices offered by mainstream supply chain members: local growers of spring mix in Sacramento, for instance, were not even willing to accept the considerably higher price that the Davis Food Co-op offered them (relative to nonlocal suppliers), unless they were experiencing a short-term glut in supply. These growers were unwilling to trade off the higher prices from the direct supply chain for the larger volumes the Co-op was able to purchase but at lower prices. Similarly, New Seasons' 2009 blueberry supplier was unwilling to lower his wholesale price beyond a 17 percent reduction of his retail price in direct markets; in 2010 New Seasons found other lower cost local producers and terminated the relationship with its long-term supplier.

Achieving certain minimum levels of production by specializing in fewer products in a given geographic location may enable local growers, as a group, to offer sufficient volumes and product consistency to benefit from agglomeration economies (i.e., benefits arising from proximity of firms from a given sector, reflected, for example in the ability to attract labor with specific skills) and from economies of scale, primarily in post-harvest activities. Lower post-harvest costs in packing, processing, and storing are therefore necessary to integrate local foods into mainstream supply chains. However, we need to know more about the economics of establishing pre- and post-harvest infrastructure targeted specifically for local food products.

Another price challenge for local foods is balancing supply and demand. Would price premiums for local foods deteriorate as they gain share in total food sales? Some farmers markets limit the number of growers marketing the same product in order to avoid creating excess supply situations that would foster price competition. Is it plausible that one large local producer could significantly impact prices for a given local product since local markets are often characterized as quite thin? We have very limited empirical evidence of the relationship between the amount of local foods supplied and their prices to answer these important questions. As discussed in chapter 8, the estimated price premiums for local foods in our study, 5 to 15 percent above the conventional prices, were lower than those commonly reported in willingness-to-pay studies. Of course, it is possible that consumers may overestimate their true interest in local products when presented with a survey, compared with their actual behavior on a shopping occasion.

A related issue is the difference in prices for local products across various retail outlets. Using statistical analyses, Boroumand (2007) found that prices in Ottawa, Canada, for local foods in retail outlets selling strictly local products were significantly higher than the prices for such products in retail outlets selling a mix of local and nonlocal products. This insight highlights the relevance of retail outlets in conveying the value of the attribute "local" to consumers and that the price premium is not only a function of the local attribute but also about the overall authenticity of the channel selling these products.

Growing Demand for Local Foods in Mainstream Channels May Be Challenging

Efforts to increase the demand for local food further face many challenges. Price, as already elaborated, is perhaps foremost among them. As with all goods, demand for local foods will depend in part on their price relative to alternatives and on the value(s) that consumers attach to the particular bundle of attributes that make up local foods. The higher prices associated with many local foods currently serve as serious barriers to price-conscious consumers striving to stretch their food budgets, particularly in difficult economic circumstances. Of course, not all consumer budgets are similarly constrained. Some consumers continue to be willing to pay more for certain attributes in local foods, among them perceived improvements in flavor and safety and for the family, community, and environment. The potential size of this latter segment has not been determined but it is unlikely to be large and has perhaps already been largely attained. Additional gains may be difficult. More research to measure the potential market size for local foods is critical to inform key private and public decision makers.

Unless there are truly unique characteristics that differentiate a local product (i.e., a local heirloom variety or a unique flavor), bringing prices in line with mainstream foods may be necessary to make significant advances in the demand for local foods. Here, for the most part, scale is key. Small producers and distributors often lack the systems, technologies, and investment capacity essential for cost efficiencies on a large scale. New supply chain configurations to address these constraints could result in lower prices and go a long way to expanding local food demand. As described, some aggregators are already collecting formerly widely dispersed lots of local foods into larger batches for more efficient transportation and processing.

Moreover, to the extent that the supply of local foods increases through expansion of existing production rather than through simple replication of independent, fragmented operations, scale economies may be achieved to provide moderation of prices. However, it is unclear whether greater production from larger growers will be accepted by some consumers as legitimately "local"; even if such foods meet strict geographical definitions, they may lack other attributes mentioned that are often associated with local. Further, it is questionable whether larger growers will be willing to adopt varieties favored by some consumers for perceived superior taste if the variety cannot withstand the rigors of long distance handling and transportation. Similarly, given scarce harvesting resources, larger growers may not be willing, or able, to add the extra labor for multiple rounds of picking that are often required to ensure peak maturity and ripeness. Finally, even if local production can increase from nearby large producers, our evidence indicates that this type of "local" may not warrant the price premiums sought by smaller producers.

Improving Entrepreneurial Skills Can Help

The potential to expand foods in local supply chains appears to exist by improving the business skills of the entrepreneur-producers. Although virtually all our case studies cite local producers who developed entrepreneurial skills on their own as they moved into post-production activities, ample room for more expertise and education exists. Agricultural producers specialize in production, which is, after all, their business, and they often have extensive training in production methods. Often, however, they lack familiarity with the myriad of additional management functions required for successful distribution and selling, and they lack the scale that can justify hiring staff with such expertise. Our case studies indicated a number of examples of post–farm gate intermediaries who are offering assistance to farmers. For example, Thousand Hills Cattle Company, a privately held marketer-aggregator in the Twin Cities metro area, annually organizes a producer meeting where it presents the latest findings on best practices in areas such as meat quality and pasture management as well as providing educational opportunities to interact with both retail and food service customers. Similarly, the meat processor in the Twin Cities case has been active in providing educational programs for direct market livestock producers, not only in Minnesota and the surrounding states but also in other parts of the country.

In many cases, mainstream channels are eager to source more local foods, but currently quality or other attributes, including management expertise, may be lacking. Providing educational efforts to address these deficiencies in local supply chains would improve access to important mainstream outlets. Our cases indicate that more training in the following areas would prove useful in strengthening the management skills of local producers: supply chain logistics, post-harvest handling, quality consistency, packaging, pricing, and delivery scheduling. In the apple case studies, Superfoods, the mainstream supermarket operator, provides considerable training to strengthen the forecasting and accounting skills of its apple growers, while Nugget Market's management coaches entrepreneurship students at local universities in Central California. The Cooperative Extension Service at land grant universities may wish to play an important part here. Whereas the traditional extension role has centered more on production and farm business management, the emergence of local foods may require a shift toward marketing, entrepreneurship, and other skills of running an integrated farm and food business.

Food Safety

Food safety requirements could constrain increases in the supply of local foods. Although many larger producers have certified food safety programs, most of the smaller producers lack such credentials.

Congress passed the Food Safety Modernization Act (S.510) in December 2010, but at the time of writing the related regulations had not yet been released by the FDA. Some policy watchers contend that it could be several years before the provisions of S.510 are enforced, because the legislation did not include any funding to FDA for the massive enforcement effort required.

Nevertheless, the "Tester amendment" specifically provided an exemption from the provisions of S.510 for qualifying smaller farms. It imposed a fourfold test to qualify for an exemption from the requirements of the legislation: (1) 50 percent or more of farm product must be sold through direct markets; (2) the direct markets have to be in the same state or within 275 miles of the farm; (3) total farm sales must be less than $500,000; and (4) the name, address, and phone number of the farmer must be provided to the customer.

Independent of any regulatory requirements, many supermarkets require their food product suppliers to be certified by a third party food safety organization. Some grocery chains, such as Superfoods in our apple case, are now working independently or with their packer-shippers and distributors to provide food safety training to local food producers. A packer-shipper in Fresno is offering training regarding the USDA Agricultural Marketing Service's Good Agricultural Practices to Hmong producers who have been supplying Asian greens to the firm; however, this effort is posing considerable language and cultural challenges.

Smaller growers could encounter financing constraints when they are required to make food safety–related improvements, such as adding restrooms. Some retailers may have an incentive to provide low-interest loans for such capital investments to smaller local growers with whom they have a strong relationship. Indeed, Superfoods not only guides the producers through the certification process but, in many cases, actually furnishes the grower with the $400 certification fee. Furthermore, Whole Foods provides up to $10 million annually in low-interest loans to small local producers. Additionally, some growers may find that they can reduce their food safety program costs by forming their own local organizations to develop and administer their food safety certification programs. Organization staff would rotate through farms within the region to achieve economies of scale in their services.

Seasonality Requires Flexible Procurement Practices

Seasonality in production is an important feature of food products in general, and arguably nowhere more important than for fresh fruits and vegetables. In some cases, fruits and vegetables even exhibit higher prices during the harvest season, as illustrated by our mainstream apple case. This is due to the availability of local varieties with unique attributes and to the sense of excitement that the harvest season generates among some consumers. But for retailers, changing suppliers is costly, since working with local farmers on a seasonal basis involves additional transaction costs. At the same time, direct delivery of local products to the store likely increases a store's operational costs, inventory control requirements, and management oversight. Indeed, for this reason, some retailers require that all local producers deliver to the wholesale warehouse, not directly to the stores.

Often these transaction costs and added inefficiencies can be minimized if the retailer sources both local and nonlocal versions of the same product from the same distributor. Some of the growers who supplied produce to one of our intermediated cases, the Davis Food Co-op, also sold other produce items to one of the Co-op's produce distributors, Nor-Cal; in turn, Nor-Cal carried both local and nonlocal versions of the same product. Coordination between mainstream packer-shippers and supermarket companies to invest in infrastructure is one potential strategy to support the supermarkets' local program(s).

An added benefit of intermediation in the distribution of local products is that mainstream packer/shippers possess technical and marketing skills that could reduce a retailer's search costs for local growers. Moreover, mainstream packer/shippers generally have the planning capacity to work with retailers in anticipation of local harvests, potentially extending the retailer's local season. However, the following conditions must be met for such investments to occur: (1) local production volumes must be large enough to keep the packer-shippers' supply chain costs low; and (2) there must be adequate demand for local products to prevent price premiums for local products from being eroded.

Our case studies suggest that in some instances, even if local products are available year-round, they are only sold and marketed as "local" during the harvest season. For example, the focal supermarket in our apple cases buys from two large local apple grower-packer-shippers year-round but markets a portion of these apples as local only from the end of August through mid-November, the harvest season. The apples identified as local during this period are merchandised in different packaging—for example, paper totes—and displayed in prime locations of the produce department. This observation suggests that the demand for local foods may be seasonal, similar to the supply of these products.

One possibility for dealing with the challenges posed by seasonality is to align local producers into strategic partnerships with counter-seasonal producers in other regions to enhance their ability to offer retailers year-round supplies and render buying more of a "one-stop shopping experience" for the supermarket procurement specialists. This requires that local growers adopt more of a marketing orientation than many currently display and that they are able to achieve minimum volumes of products with the standards set by supermarkets. Earthbound, the supplier in our mainstream spring mix case, started out as a small local supplier; it expanded by sourcing from other local growers and eventually became a national year-round supplier by sourcing from counter-seasonal suppliers in other production areas. In some cases it appears likely that large producers from far-away production areas are interested in expanding production to new locales in order to capitalize on the demand for "local" from consumers. Such strategy, however, may displease those consumers for whom the attribute "local" is associated with small nearby farms only.

Closing Comments

Future expansion of local foods may require more effective linkages between local growers and the mainstream food supply chain. To achieve this goal, two primary alternatives appear to be the most promising:

(1) connecting local growers to mainstream regional and national super-market chains; and (2) further development of natural food cooperatives. Furthermore, expanding local foods may require redefining the attribute "local" such that it allows local growers in a given region to develop partnerships with local producers in other regions. This is particularly important to overcome the limitations imposed by the low volumes and the seasonality in production that local foods face to increase their participation in mainstream channels. Moving toward a more region-alized food system is unlikely to erode the meaning of the attribute "local" among consumers, because as explained in chapter 9, distance alone is insufficient to assess the environmental performance of local and mainstream food distribution systems. Consumers may give more importance to connecting with growers and knowing the production and distribution practices employed than to the geographic proximity of production.

The mainstream supply chain appears to offer the best potential growth for local food. Although farmers markets are attracting considerable attention, direct marketing generates less than a half a percent of farm revenues and involves only about 6 percent of farms nationwide (USDA 2010). From a transportation efficiency perspective, the large load sizes involved in the distribution of products through mainstream supply chains require considerably less fuel per ton of product than the pickup trucks often used to transport foods from farms to farmers markets. Furthermore, time-strapped U.S. consumers are accustomed to conve-nience in grocery shopping; they seek out one-stop shopping, often have access to a food retailer 24 hours a day, and have year-round access to many products. Conversely, direct marketers offer a much more limited, and seasonal, product assortment and are accessible only for a limited number of hours during the week.

However, supermarkets may lack the economic incentives to locate stores in certain rural areas due to small market sizes and consequent lack of purchasing power. In such areas, the share of local foods could grow through simultaneous efforts directed at supporting the creation of farmers markets, diversifying agricultural production (e.g., perhaps toward alternative crop and animal production), and linking local pro-ducers to smaller stores in rural areas (e.g., small independent stores).

Grocery cooperatives could be a logical segment for growing local food demand. While sales in the mainstream grocery sector have been rising slowly during these difficult economic times (2.2, 1.8, and 1.0 percent growth in 2008, 2009, and 2010, respectively; Progressive Grocer 2010, 2011), the members of the National Cooperative Grocers Association reported significantly higher annual sales growth rates (9.8, 5.0, and 8.2 percent, respectively, for the same years; Pugh 2011). However, it should be noted that grocery cooperatives' sales represented only 0.2 percent of total supermarket sales in 2010. Grocery cooperatives serve as key intermediaries for local foods, since they are often willing to purchase direct from local producers. Furthermore, they are committed to providing the information flow regarding by whom, where, and how a local food item was produced that many loyal consumers of local foods desire. In effect, grocery cooperatives act as agents for their consumer-owners. However, it is unclear if grocery cooperatives can provide enough supply to gain the scale economies needed to have prices that will attract many consumers who currently shop at traditional supermarkets.

Although consumers will continue to seek out local foods through direct markets, mainstream supermarkets are likely to become the primary suppliers of local foods, just as they did eventually with organic foods. They may be able to provide the product information that consumers of local foods are seeking by capitalizing on technology and consumers' growing use of social media. To grow their supply of local foods, mainstream supermarkets must identify local producers who meet the chains' food safety and quality requirements and who are willing to engage in aggregation. Only then will they be able to offer local foods at attractive price points and reduce the transaction costs involved in their procurement and receiving activities.

Note

1. See 7 U.S.C. 1932(g)(9).

References

Boroumand, Nikoo. 2007. *An Exploration of Price Premiums and Consumer Preferences towards Locally Produced Food.* ADMN 598 Project. University of Victoria, Victoria, Canada.

Davis Food Cooperative. 2011. davisfood.coop/policies.html (accessed 2/13/2012).

Dawson, B. 2012. "Selling to Chefs, Wholesale at Farmers Markets." *Small Farm News.* University of California Small Farm Program. http://ucanr.org/blogs/blogcore/postdetail .cfm?postnum=7104 (accessed 4/6/2012).

King, R., M. Gomez and G. DiGiacomo. 2010. "Can Local Go Mainstream?" *Choices* 25(1): 19–23. choicesmagazine.org/magazine/pdf/block_44.pdf (accessed 2/13/2012).

Low, S. A., and S. Vogel. 2011. Direct and Intermediated Marketing of Local Foods in the United States. ERR-128. Washington DC: U.S. Department of Agriculture, Economic Research Service.

Mendell, D. 2009. "Cracking Quick Response Codes." *Ad Week,* January 12. adweek.com/aw /content_display/community/columns/other-columns/e3ice058ab1756ad165420 acae99d11edbd (accessed 4/15/2011).

Progressive Grocer. *Annual Report of the Grocery Industry.* 2009, 2010, 2011. April issues. Stagnito Publications.

Pugh, C. 2011. National Cooperative Grocers Association. Personal communication, March 4, 2011.

Stevenson, S. 2009. Values-Based Food Supply Chains: Red Tomato. agofthemiddle.org /pubs/rtcasestudyfinalrev.pdf (accessed 3/13/2010).

USDA. 2010. Regional Food Hubs: Linking Producers to New Markets. Know Your Farmer, Know Your Food (KNF2). Regional Food Hub Subcommittee. ams.usda.gov/AMSv1.0 /getfile?dDocName=STELPRDC5088011&acct=wdmgeninfo (accessed 2/13/2012).

Wright, R. 2010. President and CEO of Market of Choice, presentation at the 2010 National Grocery Association Conference.

11 What Role Do Public Policies and Programs Play in the Growth of Local Foods?

Michael S. Hand and Kate Clancy

As local foods have become a more popular and visible segment of the U.S. food system, there has been increased interest in public policies and programs designed to support the expansion of local foods. Much of the growth in local foods can be explained by factors outside the policy realm, such as consumer demand for product attributes and for linkages with the source of their food that are not available for products in mainstream supply chains. Yet there have been many notable attempts in the public sphere at federal, state, and local levels to encourage development of and support growth in local food supply chains.

The question of whether this is desirable from the public point of view is addressed in chapter 9, "What Does Local Deliver?" Challenges and opportunities for expanding local food product supply chains are analyzed in chapter 10, "Can Local Food Markets Expand?" Here we attempt to understand better how public policies and programs may play a role. What policies and programs are capable of affecting the development and growth of local food supply chains, and how might this occur? How do different policy options likely influence the size and structure of local food supply chains? And what can we learn from policies that may have an indirect impact on local foods? We attempt to answer these questions by examining policies and programs that are receiving attention among food system scholars, advocates, and policy makers and by drawing on experiences from the 15 supply chain case studies in this volume.

Consistent with the 15 case studies, our focus continues to be on the supply chain rather than on the individual farm, organization, or business that operates in a supply chain. When examining the potential effects of policies and programs, it is desirable, for example, to look beyond how a subsidized farmers market purchase or guaranteed loan may benefit the recipient. We also attempt to assess how different policies and programs might affect relationships among supply chain partners, which determine how products, money, and information flow between consumers and producers.

In this regard it is important to note that in none of the local supply chain cases presented in this book was there a significant interaction with any prominent public programs. The supply chain case studies all focused on private for-profit businesses that had developed their supply chains largely through their own initiative and financing or through connections with other private enterprises. We suspect that this is typical of many local food supply chains currently in existence.

However, it is possible that public policies and programs have played some role in these supply chains. For example, all the producers in the five direct market supply chains sell products in a farmers market. These markets could have received funding from the Farmers Market Promotion Program, administered by the U.S. Department of Agriculture, or may participate in federal nutrition programs that encourage fresh fruit and vegetable purchases at farmers markets. Support of this type may broaden the demand for products sold in farmers markets, particularly in areas where mainstream market outlets do not provide adequate options for fresh fruit and vegetable purchases. Also, vendors may rely on services that are collectively provided by the farmers market and are supported through public funding. But because the farmers markets were not the focus of the case studies, support of this type may exist but be unobserved in the case studies.

Selecting Policies and Programs to Examine

When it comes to the policy and politics of food and agriculture, almost any policy, law, regulation, and program can be thought of in terms that may affect local food supply chains. On the largest scale, broad issues related to international agricultural policy and trade, such as how countries design agricultural support programs to comply with trade rules, may have an impact on efforts to localize food systems (Morgan et al.

2006: ch. 2). Agricultural commodity programs, which over the decades have provided an array of subsidies, direct payments, price supports, and counter-cyclical payments to producers of major staple crops, are seen by some as a serious impediment to the development of alternative food systems (Allen 2004: 182–87).

On the other end of the spectrum, state and local food policy can be designed to affect the availability of local food in a community more directly by addressing unique needs of a particular place (Hamilton 2002). Numerous states, towns, and cities have made commitments and accommodations—both large and small—that allow local food supply chains to have a presence in their community.

We have chosen to examine several policies and programs that could have a direct impact on the types of supply chains studied in the previous chapters. Further, we consider policies and programs that are capable of having broad applicability across food products, market types, and places. Finally, although it would be most desirable to look only for clear policy connections, it is also necessary to consider less obvious and indirect impacts.

How Do Policies Affect Supply ChainStructure and Size?

It is not always clear how public policies may affect the size and structure of local food supply chains. Nor are the circumstances under which policies and programs would support growth in local foods always straightforward. For example, a program to support investments in food processing equipment may help a local producer overcome high initial startup costs that may keep some producers in local supply chains out of certain markets. Or a program could operate by creating a pool of investment funds on which local producers could draw, thereby reducing the costs and risks associated with new business development. But the usefulness of these programs depends on the entrepreneurial enterprise and skills that are necessary for a producer to expand into new markets. Programs may or may not be designed to foster such characteristics or to target participants who already possess them.

A difficulty in assessing the net impact of such programs is that we tend to observe only extant and successful supply chains. We rarely observe participants in programs that have ceased operations, nor do we observe what successful participants would have done absent the program's support.

Are Indirect Effects of Food and Agriculture Policies Important for Local Food Supply Chains?

Many of the policy areas discussed in this chapter fall into the category of directly supporting growth of local food supply chains. But other policy areas not designed to address issues specific to local foods may nonetheless have an impact. Most food moves through mainstream supply chains that do not attempt to make a connection between consumers and the source of their food. Thus it is likely that most policies will be designed to address needs in mainstream supply chains, with the potential for deleterious or constraining effects on local supply chains.

The most salient example of this type of policy relates to the regulation of safe food handling practices, which is discussed later in the chapter. Understanding these indirect effects, which may be less visible than policies and programs designed to encourage growth in local foods directly, may help provide a clearer picture of the role of policy in local food supply chains.

Local Food Policy Options and Their Potential Impact

Although policy and program interactions with the supply chain case studies are largely hidden, there are several areas of policy that could have an impact given what we did observe in the case studies. We examine examples of marketing and promotion programs, business development and loan programs, and research and extension efforts that have received much attention for their potential to support local food supply chains.

Local Foods Marketing and Promotion

Perhaps the most widely recognized programs are those that provide assistance in marketing and promotion of local products. For example, many states commit resources to the promotion of products grown or produced within the state (Onken and Bernard 2010). These programs generally seek to develop additional market outlets for local foods and to provide knowledge and technical assistance to enterprises that market local foods. For newly developing local food supply chains, particularly those that involve small enterprises and people inexperienced in food marketing businesses, marketing and promotion assistance can help overcome significant barriers to getting products from producers to consumers.

A common example of support for local food marketing and promotion is direct marketing programs. Multiple (if not most) states have sought to increase sales of food products directly from farmers to consumers by creating or subsidizing farmers markets or roadside farm stands (Hamilton 2002).

At the federal level, the Farmers Market Promotion Program (FMPP) is the primary source of marketing and promotion assistance for local direct marketing efforts. Established in 1976, the FMPP administers grants of at most $100,000 to organizations for a wide variety of activities. For example, grants may fund the startup of a farmers market and the development of marketing materials, or provide training to farmers and market managers. However, the program funding cannot be used for construction of permanent facilities.

A key component of FMPP grants in the last several years has been to increase the acceptance at farmers markets of electronic benefit transfer (EBT) for purchases with Supplemental Nutrition Assistance Program benefits (SNAP, or food stamps). Among the 81 projects funded in fiscal year 2010 by FMPP, 35 included a component to accept EBT at markets that previously could not accept EBT, or to expand accessibility at markets where EBT already exists.[1]

EBT redemption at farmers markets represents only a small portion of total purchases made with EBT, but the focus on EBT highlights an important role for programs such as FMPP. Because farmers markets tend to be temporary and periodic, they often do not involve large capital investments (such as permanent structures) that would allow them to use EBT technology. Further, any one farmer selling at a market likely does not have an incentive to invest in the ability to accept EBT. FMPP grants have the ability to concentrate the necessary resources on an investment that no one farmer could make but can benefit all farmers in the market by broadening the market's customer base and creating a market outlet that might not exist without public funding. This effect may be largest when prices at farmers markets may be comparable or lower than in supermarkets (e.g., during certain parts of the growing season). The ability to use EBT at lower-cost farmers markets could give recipients additional flexibility in stretching their food dollars. However, to our knowledge there is no existing research that examines the importance of EBT acceptance for the success and growth of farmers markets.

Business Development Programs

Whereas marketing and promotion programs seek to make connections between consumers and producers in the food marketplace, there has been significant interest in helping nascent enterprises develop into viable local food businesses and supply chains. Business development programs may provide low-cost loans for infrastructure development, or may provide assistance in developing business and marketing plans for enterprises entering new markets.

The value of this type of assistance is that it can reduce some of the costs of certain business startup activities and can reduce the risk of starting or expanding a business. In some cases this may occur through subsidized loans, grants from state or federal agencies, or cooperative agreements through nonprofit organizations (which may themselves be funded in part by state or federal money). For small enterprises, particularly those serving smaller and rural markets, business development programs may represent one of the few options for obtaining working and physical capital. Also, community food enterprises, such as food banks, may benefit from business development programs if they have a difficult time obtaining funding from private sources.

The Business and Industry (B&I) Guaranteed Loan program, administered by the USDA Rural Development Agency, is one example of a program that provides financing support for businesses and nonprofit and community organizations in rural areas. In this case, B&I loans must be for projects that meet certain criteria, such as improving economic or environmental conditions in the community or increasing use of renewable energy. B&I loans may fit well with some local food enterprises that emphasize locally produced food as part of a sustainable agriculture initiative.

The B&I loan program is unique in that it specifically defines the concept of "local foods" for the purposes of providing financing to local food projects and sets aside 5 percent of funds for enterprises that are engaged in local food supply chains.[2] Also, recipients of the loans who sell through retail and institutional outlets are required (through agreements with the retail and institutional buyers) to inform customers that the food is locally produced. This unique clause essentially mandates, for the purpose of a portion of loan funding, that the products are distributed in local food supply chains; that is, where information about a product's origin is conveyed to consumers.

A key characteristic of the B&I loan program is that it is primarily designed to support enterprises in rural areas. Only in certain circumstances may a recipient be located in an urban area (defined as a town or city of 50,000 people or more, and its surrounding urbanized area). Even with these exceptions, the recipient must show that the primary beneficiaries will be located in rural areas.

The focus on rural areas in the B&I program may present a challenge for widespread applicability of the program for the purposes of supporting local food supply chains. Growth in markets for local foods has thus far been primarily an urban phenomenon. Urban areas provide the largest markets and most varied demand, which allows new businesses and enterprises of varied size to find viable market niches. However, many of the farms and businesses that supply local food markets in urban areas are located in rural areas. Because cities provide the largest market opportunity for local foods, rules that restrict participation in urban or peri-urban areas may exclude enterprises that have a high probability of growth and success.

The case studies provide evidence that both rural- and urban-based businesses in local food supply chains can be successful. Several of the enterprises in the supply chain case studies are located in rural areas and would likely satisfy both the "rural" and "local" definitions in the B&I loan program. Yet urban linkages may be a key success factor. All of the local supply chain cases have significant economic linkages to enterprises based in urban areas and rely on access to larger urban markets.

Striking a balance between the desire to support enterprises in rural areas and the ability of those enterprises to access markets that offer a high chance of success may require additional flexibility. This could occur by funding loans that support enterprises in local food supply chains that are based in urban areas but partner with other businesses (like producers and processors) that are located in rural areas and sell products in urban markets. This type of expansion in eligibility would move away from the traditional direct support of businesses in rural .areas, yet it recognizes that economic linkages in local food supply chains (and mainstream chains as well) are not clearly delineated by rural vs. urban distinctions.

Community Access and Health

A major use of policy and programs in support of local foods has been to encourage greater access to healthy and affordable food, particularly in areas not well served by mainstream supply chains. Policy and program support could plausibly encourage local producers and businesses to fill a gap in healthy food availability for some communities, and this type of support could ultimately spur the development of robust local food supply chains.

Prime examples of support to short or direct market local food supply chains to encourage healthier community food access are the Women, Infants, and Children (WIC) Farmers Market Nutrition Program and the Seniors Farmers Market Nutrition Program. These programs provide grants to states to give participants coupons worth $10–$30 per market season to purchase fresh produce at farmers markets. In short, purchases of fruits and vegetables by eligible participants are subsidized, which tends to increase demand for those food products relative to products that are not subsidized. If the resources and demand for nutrition program–eligible products increases enough, this could act as a "flypaper effect" that provides an incentive for the shorter local food supply chains to develop.

Programs can also operate on the supply side. The Healthy Urban Food Enterprise Development program (HUFED), funded until fiscal year 2013, supported local food supply chain development by reducing barriers that may previously have kept food enterprises out of certain communities. Among the first-year HUFED projects funded by USDA was one allowing a seed and mill company to sell staple food grain products in local markets where lower-income consumers shop. The program was designed to overcome barriers posed by labeling requirements and other startup costs. Targeting specific supply-side impediments in underserved areas could create an incentive for local food supply chain development to take advantage of a new market opportunity.

Programs such as these may provide either a demand or supply "boost" where local food supply chains operate or could operate in the future. The question is whether it is possible to remove all the relevant barriers that prevent supply chains from developing in underserved areas. For example, mainstream supply chains may not serve certain communities well because economic factors create a barrier to success (e.g., a low-income population in sparsely populated areas may be difficult to serve profitably).

Research

The rapid growth in local food markets has resulted in numerous business, marketing, and nonprofit experiments as enterprises attempt to achieve programmatic goals and succeed in a new market sector. As with any new or growing market sector, there is a great demand for understanding what technologies, systems, and business practices work well in different situations. The wide variety of existing supply chain arrangements for local foods suggests that what works in one situation may not be applicable in another; it also suggests that there may be many business models and supply chain arrangements that work better than others.

Research and development programs may be able to support local food supply chains on a fundamental level. Publicly funded research and extension programs to disseminate findings and applicable technologies, may allow local (or mainstream) supply chain partners to draw on significant investments in research and development that most enterprises would not be able to engage in on their own. This may be particularly important for enterprises that begin operations with small initial investments and tight operating budgets; starting with the right technologies and business model can help these operations avoid costly mistakes and improve chances of success.

The Federal-State Marketing Improvement Program (FSMIP) is an example of a program with a significant research component that has made investments in local food supply chain research. FSMIP and other research and extension programs (such as the Sustainable Agriculture

Research and Extension program) often leverage resources already in place through state departments of agriculture, land grant universities, and agriculture experiment stations. These resources have a long history of providing research and development benefits for food and agriculture industries and are likely well positioned to provide support for local food supply chains.

FSMIP provides federal matching funds to state agencies to explore new markets for food and agricultural products, and several projects contain a research component that could develop knowledge and technologies appropriate for local food supply chains. For example, a project funded in fiscal year 2010 in New Jersey is supporting the development and launch of New Jersey grown and processed value-added products that meet the nutritional and cost requirements of the National School Lunch Program.

In another example, a project in Massachusetts sought to examine the impact on vendors and beneficiaries of the new WIC voucher programs at farmers markets. The program allows WIC participants to receive cash-equivalent vouchers for fruit and vegetable purchases from farmers market vendors. This type of research could help fill some gaps in the understanding of how nutrition programs are related to local food supply chains, how nutrition programs affect purchases of fresh fruit and vegetables, and what the net effect of these programs is on the demand for local food products.

*Issues with Economies of Size and
the Case of Food Safety Requirements*

Thus far we have primarily focused the discussion on policy areas that may directly affect local food supply chains. But because the vast majority of food and agriculture policy is concerned with mainstream supply chains, it is reasonable to expect that some nationwide food policies will have unintended effects on local supply chains. For example, we have in the previous sections alluded to a connection between a policy's impact on smaller enterprises and its consequences for local supply chains. Enterprises in local supply chains are not, by definition, small. But we do observe (in the case studies and elsewhere) that local food supply chains are smaller in scale, and that the enterprises making up those supply chains tend to be smaller than their mainstream counterparts. Thus it is possible that policies with unintended effects on small enterprises disproportionately affect the growth and development of local supply chains.

Economies of size—where per unit costs of compliance with food safety programs (e.g., per acre on a farm or per pound of product) are lower or benefits provided by complying farms are higher for larger enterprises—may result in policies that unequally affect smaller enterprises in local food supply chains. A recent example where size and compliance costs received attention was the Food Safety Modernization Act of 2011.[3] Agricultural and food handling practices that reduce the risk of food-borne illness are widely believed to be more costly for smaller producers to adopt (Hardesty and Kusunose 2009), which results in concern that the law would financially burden smaller enterprises. As a direct response to this concern, the act included an exemption for small and local producers from its provisions.[4]

In addition to recognizing the potentially uneven impact of compliance, it is possible that this type of exemption is justified on efficiency grounds; if compliance costs are not scale neutral, requiring the largest operations to comply may be a cost-effective way to provide the greatest amount of food safety protection. The additional cost of bringing small producers (whether they are in local supply chains or not) into compliance may not be justified in terms of additional food safety protection. This is not to say that small enterprises or those in local food supply chains are inherently safer. And in fact, the Food Safety Modernization Act allows for farms linked to or at high risk of a food-borne illness outbreak to lose their exemption. But in terms of the benefits from reduced risks from food-borne pathogens, a cost-effective policy would encourage enterprises with the lowest compliance costs and largest potential benefits to participate. An analogous problem is the choice of the appropriate food safety technology that balances marginal benefits and costs (see Jensen et al. 1998 for an example applied to food safety technology in the meat sector); additional investments in more costly risk reduction may increase safety benefits, but at a higher cost that reduces efficiency and imposes burdens on smaller producers.

An alternative to policies with exemptions may be programs with voluntary participation, such as the National Leafy Greens Marketing

Agreement.[5] Voluntary participation may be desirable because those supply chain partners with the lowest costs relative to the benefits of participation may be encouraged to participate. In the case of food safety practices, producers and food handlers may have an incentive to participate in voluntary programs as more retailers in mainstream food supply chains require additional, and possibly more stringent, safe food handling practices and third party certification. Or voluntary programs can provide technical assistance that reduces the cost of adopting practices that may not be specifically required by retailers or by law.

Are voluntary programs scale-neutral? The answer to this question likely depends more on the types of practices involved in the program more than on whether participation is voluntary. For voluntary food safety programs, it may be that some enterprises in local supply chains may be able to participate at lower costs, and the potential to access mainstream retail outlets would provide an incentive to participate. Others may be pushed into supply chain arrangements where these requirements are less important, such as direct marketing. If economies of size determine costs of compliance and participation, then the larger enterprises would be more likely to participate and move into supply chains involving mainstream outlets, and smaller ones would have an incentive to seek other local supply chain arrangements. But these possibilities have not yet been borne out by careful study, and more research is needed to consider the potential impact of scale on voluntary programs affecting local food supply chains.

There may be room for other policies that both overcome economy of scale problems and encourage smaller-scale enterprises to adopt additional safe food handling practices. These could be some form of the programs already discussed—business development loans or grants, research on appropriate scale for certain practices, and technical assistance. These types of policies may be desirable if enterprises that are exempt from food safety regulation can, in aggregate, provide significant compliance benefits. For example, it may be that providing technical assistance and cost sharing for capital investments is a cost-effective way to encourage safe food handling practices among a large group of small and midsized enterprises (regardless of whether they participate in local or mainstream supply chains).

Concluding Thoughts

We have described a few federal policies that have the potential to support local food supply chains. But there are myriad other policies at the federal, state, and local levels that have the same goal. Among the laws and ordinances passed are local food preferences or mandates for institutional procurement, healthy food initiatives in institutions, food financing initiatives, regional food development, and many others.

All programs have the potential to support local food supply chains in a meaningful way, but each also has potential shortcomings that may limit its impact and reduce cost-effective use of public funds. This highlights a larger issue of examining the potential impact of public policies and programs: how do we determine which policies are effective in different situations, and how do we identify instances when a program is not providing the intended benefits? Many programs are not evaluated and researched in a way that allows for valid conclusions to be made about their efficacy. In general, such evaluation would compare outcomes for program participants or beneficiary communities to an appropriate counterfactual group. That is, there is a need to compare program outcomes carefully to what would have happened in absence of the program. Additional emphasis on examining the impact of these programs can facilitate improved policy design and speed knowledge sharing among supply chain members and program participants.

For example, there is a need for study of the impact of the farmers market promotion program (and related other programs) on the price and availability of locally produced food, consumer spending and purchasing habits, and producer participation in local food supply chains. There is even less knowledge about the impact and effectiveness of programs that support longer intermediated chains like the ones studied in this volume. The most salient programs and policies related to local foods tend to target short supply chains and direct marketing. Yet if growth potential for local food supply chains is greater for local foods that move through mainstream channels, then we need to know more about how existing programs affect longer supply chains. Another pressing question is whether program support for local food supply chains is adequate to circumvent the barriers to success that have prevented other supply chains from developing in the absence of the program.

There is also a need to understand better the public benefits and costs of programs that can affect growth in local food markets. The primary objective of several of the policies discussed in this essay may not be growth in local foods but the provision of other public benefits. For example, improvements in public health and nutrition are the goal of the farmers market nutrition programs and HUFED; it is possible that local foods can be a vehicle for achieving that goal, but growing local food markets is not itself a good measure of success in these programs. Whether or not growth in local foods helps to meet program objectives, and whether these benefits are worth the cost, have yet to be examined for most programs.

Looking forward, there are several policy interventions that have the potential to affect the development of local food supply chains. One is to target the development of state or regional food plans for meeting food system goals, such as health, food security, food safety, and environmental sustainability. These plans might account for energy, water, and climate change effects as well as current and future production and processing capacity in the region. Under a state or regional focus, plans could be prepared by experts from across all food sectors of the food system under the direction, for example, of a state governor's office or commissioners of agriculture from multiple neighboring states. We do not know of a prototype yet, but most of the tools for such an exercise are in place. Assessments may be designed to provide guidance on the amount of food production, processing, and distribution that can realistically be accomplished at different geographic scales and may examine the infrastructure, financing, and incentives that might be necessary to achieve a range of food needs. A risk of this type of program is that local food systems, regardless of their design and operation, will be assumed to achieve the plan's food system goals. It is more appropriate to examine how different characteristics of supply chains contribute to meeting food system goals and to encourage the development of those characteristics through the food plan.

Potentially useful and complementary institutions are food policy councils, many convened at a local level, which are addressing food system policy issues in many communities (Feenstra 1997). Some councils assess policy and programmatic needs on interrelated issues such as health and nutrition, economic development, food security, and environmental quality. These councils tend not to focus on individual policy issues in isolation from each other and may include roles for education, networking, and facilitation as much as working toward specific policy changes (Schiff 2008).Another complementary program area is funding for technical assistance and access to capital. These types of assistance fit well within the historical mission of public funding for food and agriculture programs that develop new markets, diffuse knowledge, and accelerate food supply chain growth. Finally, it may be useful to determine how existing policies can be adapted and combined to support multiple types of food supply chains, such as longer intermediated supply chains and supply chains that convey food values from farms to consumers. As public policy concerns are increasingly addressed through a focus on the food system, we think it useful to expand the thinking about local foods beyond direct marketing and short supply chains.

Notes

1. FMPP grant awardees by fiscal year are listed on the USDA Agricultural Marketing Service web site, ams.usda.gov (accessed 10/3/2011).
2. See 7 U.S.C. 1932(g)(9): "The term 'locally or regionally produced agricultural food product' means any agricultural food product that is raised, produced, and distributed in, (I) the locality or region in which the final product is marketed, so that the total distance that the product is transported is less than 400 miles from the origin of the product; or (II) the State in which the product is produced." The program also gives priority to enterprises serving underserved communities with low access to fresh fruits and vegetables and high rates of food insecurity.
3. Public Law 111–353, January 4, 2011.
4. The act does not directly define local food producers but defines "qualified end-users" of food. To qualify for the exemption, a farm must sell the majority of its food to qualified end users, who are either the final consumers or restaurants and stores within the same state or within 275 miles of the farm. Specific definitions related to the exemption will be determined as a part of the regulatory rule-making process.
5. For details see "Proposed National Marketing Agreement Regulating Leafy Green Vegetables," *Federal Register*, 76 FR 24292, 6 CFR 970. Rulemaking for the agreement was terminated December 5, 2013.

References

Allen, P. 2004. *Together at the Table: Sustainability and Sustenance in the American Agrifood System*. University Park: Pennsylvania State University Press.

Feenstra, G. W. 1997. "Local Food Systems and Sustainable Communities." *American Journal of Alternative Agriculture* 12(1): 28–36.

Hamilton, N. D. 2002. "Putting a Face on Our Local Food: How State and Local Food Policies Can Promote the New Agriculture." *Drake Journal of Agricultural Law* 7(2): 407–54.

Hardesty, S. D., and Y. Kusunose. 2009. *Growers' Compliance Costs for the Leafy Greens Marketing Agreement and Other Food Safety Programs.* Research Brief, UC Small Farm Program. sfp.ucdavis.edu/Docs/leafygreens.pdf (accessed 2/11/2012).

Jensen, H. H., L. J. Unnevehr, and M. I. Gómez. 1998. "Costs of Improving Food Safety in the Meat Sector." *Journal of Agricultural and Applied Economics* 30(1): 83–94.

Morgan, K., T. Marsden, and J. Murdoch. 2006. *Worlds of Food: Place, Power, and Provenance in the Food Chain.* Oxford: Oxford University Press.

Onken, K. A., and J. C. Bernard. 2010. "Catching the 'Local' Bug: A Look at State Agricultural Marketing Programs." *Choices* 25(1).

Schiff, R. 2008. "The Role of Food Policy Councils in Developing Sustainable Food Systems." *Journal of Hunger and Environmental Nutrition* 3(2–3): 206–28.

12 A Look to the Future

Robert P. King, Miguel I. Gómez, and Michael S. Hand

Our contemporary food system is the product of an evolutionary process that began in the late nineteenth century with the development of new technologies for food processing and preservation, new modes of transportation and communication, and new forms of business organization. National and international brands, large and highly specialized production systems, year-round availability of a wide array of fresh products, the supermarket, and mass media food advertising are all twentieth-century innovations. The early twenty-first-century local food movement is at once a part of that evolutionary process and a reaction against it.

The allure of local food products stems from their potential to deliver fresh and healthy food, direct connections with producers, personal assurance of sustainable production and distribution practices, economic development, and the reinvigoration of cultural connections and social ties. The case studies in this volume demonstrate that local food products *can* deliver all these things, but they also show that local food supply chains are not always the most effective mechanisms for meeting these desired food system outcomes.

As we look to the future, we believe that the local food movement will continue to influence and complement the broader food system around three important issues: transparency, sustainability, and sense of community:

1. Growth in demand for local food products has highlighted the importance of *transparency* in the food system. The focus in local

346

food supply chains on direct links with food producers and on the conveyance of information about where, by whom, and how food is produced has focused consumer attention on understanding the provenance of the food they purchase. This carries over to mainstream grocery and food service channels, where retailers are making efforts to provide more information about the products they sell and to create meaningful bonds between consumers and more distant producer groups.

2. The local food movement has focused on fundamental aspects of the economic, environmental, and social *sustainability* of the food system. On the one hand, local food advocates have drawn attention to some unsustainable practices in mainstream supply chains. On the other hand, the case studies in this volume demonstrate that some local food supply chains fall short in ensuring economic viability for producers and other chain participants and in making most effective use of environmentally sensitive resources. Independent of the outcomes of local and mainstream supply chains, the emergence of the local food movement has helped bring to the forefront important questions about the environmental impacts of food production and distribution, the fair distribution of costs and returns across the supply chain, fair treatment of workers, and food safety.

3. Food is a key element of any culture, and eating is often a social activity. The local food movement has helped strengthen the *sense of community* in communities large and small. Ironically, in many communities place-based interest in food developed first with a focus on ethnic foods. The growth of markets for local foods has encouraged consumers to learn more about foods originating in their own locale. This creates a stronger sense of connection to the local community and establishes a context within which people can understand and appreciate their diverse cultural heritages.

While the volume of food moving through local food supply chains is relatively small, the local food movement has had significant impacts on discourse about the food system and on the competitive landscape in food retailing. We believe consumers' definitions of *local* will continue to evolve, as will the place of local products within mainstream grocery and food service supply chains. We also believe that there will continue to be attractive opportunities for local food entrepreneurs and that local food advocates will continue to raise important questions about the structure and performance of our food system.

CONTRIBUTORS

KATE CLANCY is a senior fellow at the Minnesota Institute for Sustainable Agriculture at the University of Minnesota. She has research and program interests in regional food systems, agriculture of the middle, and agriculture and food policy.

GIGI DIGIACOMO is a research fellow in the Department of Applied Economics at the University of Minnesota. Her research addresses farm business planning, organic commodity marketing, and food system performance.

MIGUEL I. GÓMEZ is an associate professor in the Charles H. Dyson School of Applied Economics and Management at Cornell University. His research program focuses on food marketing and distribution, sustainability of food systems, and supply chain and price transmission analysis.

MICHAEL S. HAND is an economist with the U.S. Forest Service, Rocky Mountain Research Station, in Missoula, Montana. He conducts research on wildfire suppression costs, risk management incentives for wildfire management, and values of public forest ecosystem services.

SHERMAIN D. HARDESTY is an extension specialist in the Department of Agricultural and Resource Economics at the University of California, Davis. Her research and outreach focus on local food systems, small farms, and collaborative business models.

JEFFREY HORWICH is a former graduate student in the Department of Applied Economics at the University of Minnesota.

ROBERT P. KING is a professor in the Department of Applied Economics at the University of Minnesota. His research and teaching focus on local food systems, organic agriculture, supply chain management, and organization design.

LARRY LEV is a professor and extension specialist in the Department of Applied Economics at Oregon State University. His agricultural marketing efforts focus on local and regional food systems and on the development of innovative research and outreach approaches.

EDWARD W. MCLAUGHLIN is the Robert Tobin Professor of Marketing in the Dyson School of Applied Economics and Management at Cornell University. His research and teaching focus on food and produce marketing and on retailing strategy.

GERALD F. ORTMANN is a professor of agricultural economics in the School of Agricultural, Earth and Environmental Sciences at the University of KwaZulu-Natal in South Africa. His main research interests are in production economics, agricultural policy, and institutional economics.

KRISTEN S. PARK is an extension associate in the Dyson School of Applied Economics and Management at Cornell University. She conducts research on food marketing issues, with an emphasis on the fresh produce distribution system.

INDEX

Page numbers in italic indicate figures and tables.

diversification: in marketing channels, 151, 195–96, 203, 211, 216; in product offerings, 243–44
Durham, Catherine A., 14

Earthbound Farm: and *E. coli* outbreak, 140, 165; expansion prospects for, 140–41; food miles and fuel use of, 138–39, 161–62; marketing and retail for, 137–38, *139*; prices and pricing for, 160, *161*, 168; processing flow chart of, *136*; product flow chart of, *133, 155*; production and processing for, 132–35; size and growth comparisons of, 169–70; spring mix ingredients of, 129; and Veritable Vegetable, 159
Eastwood, D. B., 275
EBT (electronic benefit transfer) benefits, 56, 61, 334–35
economies of scale: and aggregation, 316–17; and energy use, 22, 225; and local producers, 320, 322; and public policies, 339–41
economy. *See* local economies
educational programming: in cooperatives, 315; for producers, 214, 323. *See also* information flow
Edwards-Jones, G., 302
Eldon, Jim, 142–44, 145–52, 170. *See also* Fiddler's Green Farm
electronic benefit transfer (EBT) benefits. *See* EBT (electronic benefit transfer) benefits
employment and labor: in apple case studies, 49, 54–55, 59, 64, 65; in beef case studies, 182, 193, 209, 211, 215; in blueberry case studies, 90–91, 92, 100, 108, 120–21; as expansion constraint, 61–62; and local economy impacts, 21; in milk case studies, 238, 246, 247, 249–50; in spring

mix case studies, 140, 142, 149, 156, 164–65
energy use, 22, 298–300. *See also* food miles; fuel usage
ethylene storage technology, 65
every day low price (EDLP) strategy, 46
expansion. *See* growth and expansion

face-to-face chains (SFSC chain type), 15. *See also* direct market supply chains
Fairchild, Jeff, 111–12, 112–13
farmers. *See* producers
farmer cooperatives. *See* producer cooperatives
Farmers' Market Federation of New York (FMFNY) study, 60
Farmers Market Health Bucks program, 61
Farmers Market Promotion Program (FMPP), 334–35
farmers markets: and aggregation, 316, 318; Davis Farmers Market, 144–45, *149*, 168; effects of policies and programs on, 331; food access programs in, 61, 62, 334–35, 337; and food security, 303–4; guidelines and rules at, 103–4, 196; and healthy eating habits, 302–3; as infrastructure, 19; and local food expansion, 62; Marin Farmers Market, 144, 145, 146–47, *149*, 168; in milk case studies, 239; Mill City Farmers Market, 196–97, 200, 201; perceived benefits of, 56, 60; prices and pricing in, 279–83, *285*; product availability data, 268–74; Santa Monica Farmers Market, 318; vendor numbers at, 320–21. *See also* Central New York Farmers Market
farm stands, 39, 64, 102, 104, 318

GAP (Good Agricultural Practices)
program, 47, 51, 96
Giraud, K. L., 275
Goodman, Drew, 132
Goodman, Myra, 132
GPS1 (SuperFoods' apple supplier),
41–43, 46–49, 51–52, 77–80, 84n7
GPS2 (SuperFoods' apple supplier),
42, 44–46, 51–52
GPS3 (SuperFoods' apple supplier),
43, 44–46, 47, 48, 77–80, 82, 84n7
GPS4 (SuperFoods' apple supplier),
43, 44–46
grants, 334–35, 337
"grass-fed" attribute, 180, 199, 202,
204, 276, *286*, 287–88, 307n11
Griffin, M. R., 280
grocery cooperatives. *See* food
cooperatives
growers. *See* producers
growth and expansion: by aggregation,
316–19; in apple case studies, 56–57,
61–62, 75, 76; in beef case studies,
189–90, 201–2, 203, 215–17, 221; in
blueberry case studies, 97, 106, 115,
119; and food safety regulations,
324–25; importance of information
flow during, 314–16; of local food
chains, 313–14; of local food products, 321–22, 326–28; in milk case
studies, 234, 242, 252–53, 256–57;
and product availability, 325–26;
in spring mix case studies, 140–41,
150–51, 165, 168–70; through producer education, 322–23; types of, 20
Gunderson, M., 303

H-2A certification program, 54
hand harvesting, 101, 130
Hannibal (NY) School District, 62–63,
65–66; business relationships of, 68,
74; community impacts of, 80–81;

and DOD fresh fruit program, 71;
economic impacts of, 79; key lessons
from, 72–73; and marketing and
distribution, 66–68; performance of,
68, 76–81; product flow chart of, *64*;
product quality and safety standards
of, 76; and promoting healthy eating
habits, 69–70; size and growth of,
71–72, 75–76; transportation and
fuel use of, 69, 80, *82*
Hansmann, Henry, 18
Hart, Oliver, 18
harvesting systems: in spring mix case
studies, 133–34, 143–44, 156. *See also*
production systems
health benefits, 60, 87, 301–3, 302–3
healthy food access programs, 337–38
Healthy Urban Food Enterprise
Development (HUFED) program,
337–38
Hine, S., 275
Hinson, R. A., 275
Holmes, Paul, 155, 164, 165–66
home delivery, 236, 238–39
"hormone-free" attribute, 187, 213, 234,
254
hospital farm stands, 102, 104
Hurst, Mark, 89, 97
Hurst's Berry Farms: business relationships of, 91; comparisons of
performance of, 119–21; comparisons of structure of, 116–18; food
miles and fuel use of, 96–97, *122*;
food safety requirements of, 95–96;
growth potential of, 97; impacts of,
97; and packing and distribution,
91–94; product flow chart of, *90*; and
production, 89–91, 124n2, 124n6

Imperial County (CA), 130, *131*
income and wages: in apple case studies, 79; in beef case studies, 189, 193,

215; in blueberry case studies, 105, 109, 120–21; local economic benefits from, 21; in milk case studies, 238, 246, 247, 249–50, 259. *See also* employment and labor; revenues and sales

information flow: during aggregation, 10–11, 318–19; in apple case studies, 45–46, 73, 74; in beef case studies, 185, 187, 194, 209–10, 213; in blueberry case studies, 93, 95, 96, 102, 110, 111, 116–17; and the definition of local, 9, 15; and growth of local food, 314–16, 328; in milk case studies, 233, 255; in spring mix case studies, 133, 145, 147–48, 160, 164, 167; transparency in, 346–47

intermediated supply chains: information flow in, 10–11; potential role in growth of local foods, 325–26; producers' net prices in, 297–98; and relation to revenue and farm size, 296–98. *See also specific supply chains*

Isengildina-Massa, O., 275

J & B Wholesale Distributing, *184,* 185–87, 188

Jim Smith Farm, 52–53; business relationships of, 74; comparisons of performance of, 76–81; comparisons of size and growth of, 75–76; expansion prospects for, 61–62; food miles and fuel use of, 59, 80, *82;* impact of farmers market on, 62; impacts of, 59–61, 79, 80–81; and information sharing on product origin, 73; marketing and distribution for, 55–57; product flow chart of, *54;* production of, 53–55; product quality and safety standards of, 76

Kaiser-Permanente, 104

King, Robert P., 14, 316

Kowalski, Jim, 182

Kowalski, Mary Anne, 182

Kowalski's Markets: and distribution, 185–87; food miles and fuel usage of, 188; impacts of, 182, 189; key lessons from, 224–25; marketing and retail for, 187–88, 212–13; product flow chart of, *184;* production and processing for, 183–85; size and growth of, 189–90, 219–21; structure and performance of, 182–83, 219, *220,* 221–24

labeling. *See* product attributes

labor. *See* employment and labor

Lancaster Organic Farmers Cooperative (LOFCO), 246, 247, 252, 256

Land Stewardship Project, 214

Langton, Brenda, 198–99

large producers, 295–97, 322

Larsen, Jim, 206–8, 214

Larson, A., 301–2

Leafy Greens Marketing Agreement, 129, 140, 166, 169

Lempert, Phil, 107

Lighthouse Farm Network, 150

livestock. *See* beef; beef supply chains; milk; milk supply chains

loans, 335–37

local, defined, 14–16, 30n1, 289, 289n4, 336, 344n2

"local" attribute: and food security, 304; need for redefining, 314, 327; and price premiums, 21, 77, 79, 119–20, 257, 276–79, 287–89; and seasonal demand, 326; in willingness-to-pay studies, 275

local communities: civic engagement in, 22; food's connection to, 347; impacts in apple case studies, 49,

Marsden, Terry, 15

Martinez, S., 301

Maryland and Virginia Co-op: and aggregation, 316, 317; expansion potential of, 234; key lessons from, 235; local impacts of, 233–34; policy issues affecting, 234–35; product flow chart of, 232; and South Mountain Creamery, 239; structure and performance of, 231–33, 255, 257–59

Maryland Division of Milk Control, 243

Matthews, H. S., 300

May, Steve, 186

McCann, N., 303–4

meat. See beef; beef supply chains

median prices (farmers markets), 281–83

Melendez, Hector, 155

mesclun. See spring mix supply chains

milk: factors influencing prices of, 284, 285, 288; and price premiums for "local" attribute, 276–79; and product availability, 273–74

milk supply chains: benefits of studying, 8–9; comparisons of performance of, 257–59; comparisons of size and growth of, 256–57; comparisons of structure of, 255–56; key lessons from, 243–44, 253–54, 259–62; and local production area, 228–29; producers' net prices in, 297–98; production statistics for, 230–31; and product overview, 230. See also Maryland and Virginia Co-op; South Mountain Creamery; Trickling Springs Creamery

Mill City Farmers Market, 196–97, 200, 201

minimum prices (farmers markets), 281–82

Minneapolis (MN). See Twin Cities (MN)

MOM's Organic Market: food miles and fuel usage of, 249; growth potential of, 252–53; key lessons from, 253–54; local economic impacts of, 249–51; and marketing and distribution, 247–49; product flow chart of, 246; structure of, 244–45

Monterey County (CA), 130, 131

National Leafy Greens Marketing Agreement, 129, 140, 166, 169

"natural" attribute, 286, 287–88

natural foods stores, 268–74, 285

Natural Selection Foods. See Earthbound Farm

Nature's Fountain Farm: comparison of performance of, 119–21; comparisons of structure of, 116–18; food miles and fuel use of, 114, 122, 124n12; and food safety, 114; growth potential of, 115, 119; and marketing and distribution, 109–10, 112–14; product flow chart of, 109; and production, 108–9; structure and performance of, 108

New Seasons Market: comparison of performance of, 119–21; comparisons of structure of, 116–18; food miles and fuel use of, 122, 124n12; and food safety, 114; impacts of, 115; and marketing and distribution, 110–14; product flow chart of, 109; structure and performance of, 107–8

New York (state), 35–39, 59

New York (state) apple suppliers, 40–43, 44, 45–49, 51–52

New York Department of Agriculture and Markets, 39, 71

New York Harvest for New York Kids Fest, 70

nonlocal, defined, 16, 30n2

Nor-Cal Produce: comparisons of structure of, 167–68; and Davis Food Co-op, 154, 159–60; and distribution, 135–37; expansion prospects of, 140–41; food miles and fuel use of, 138–39, 161–62; impacts of, 140, 174; and marketing and retail, 137–38; product flow chart of, *133, 155*

Norwood, F. B., 293

Nugget Markets: comparisons of performance of, 170–74; comparisons of size and growth of, 169–70; comparisons of structure of, 167–68; and distribution, 135–37; expansion prospects for, 140–41; food miles and fuel use of, 138–39; impacts of, 139–40, 174; information flow in, 319; and marketing and retail, 137–38, *139*; product flow chart of, *133*; structure and performance of, 131–32

nutrition assistance programs: evaluation of, 343; and food security, 303–4; impact of farmers markets on, 62; Seniors Farmers Market Nutrition Program, 337; Supplemental Nutrition Assistance Program (SNAP), 56, 61, 334–35, 339; Women, Infants, and Children (WIC) Program, 56, 337, 339

Oase, Boyd, 183, 186, 187

Odenthal, Laura, 193

Odenthal, Randy, 193

Odenthal Meats, 193–95, 201

Ontario Orchards: business relationships of, 68; ordering and distribution for, 66–68; prices and revenues of, 68; product flow chart

of, *64*; and production, 63–65; and transportation, 69

optimal ownership theory, 17

Organically Grown Company, 108, 112, 114

organic apples, 38, 268, 283

"organic" attribute, 276, 278–79, 287, 288

organic blueberries, 88, 271, 280–82

organic milk, 244

organic spring mix, 271, 280–82

Orr, R. H., 275

Ostrom, Marcia, 14

"package size" attribute, 88, 270–71, 278, 283, *285*, 287

packaging systems: in blueberry case studies, 91–94; in milk case studies, 237–38; in spring mix case studies, 143–44. *See also* processing systems

"percent lean" attribute, 276, *285*, 287, 288

performance. *See* supply chain performance

Pirog, Rich, 21, 301–2, 303–4

policies and programs: access and health, 337–38; business development, 335–37; DOD Fresh Fruit and Vegetable Program, 63, 66, 70–71, 73, 76; impacts of, 11–12, 73, 330–33, 342–44; indirect effects of, 333; as infrastructure, 19; marketing and promotion of, 69–70, 333–35; renewed interest in, 5; research and development for, 338–39. *See also* food safety

Portland (OR): demographics of, 85–87; factors influencing pricing in, *278, 284,* 287–88; farmers market pricing in, 280–82; and local production area defined, 16; product availability in, 268–74

Ramsay Highlander (manufacturer), 134

Red Tomato (marketing organization), 317–18

regional food plans, 343

regulations. *See* food safety

relationships among supply chain members: in apple case studies, 68, 73–74; in beef case studies, 186, 194, 196, 206, 210, 216, 218, 219; in blueberry case studies, 91, 111–12, 117–18; in milk case studies, 254, 255; and pricing, 17–18; in spring mix case studies, 141, 161, 167

relationships with consumers: in apple case studies, 56–57; in beef case studies, 200–201, 203; in milk case studies, 239, 241; in spring mix case studies, 147–48

replication, growth through, 20

research and development programs, 338–39

restaurant market channel, 145–46, *149*, 198–99

Restoration Raw Pet Food, 216

revenues and sales: in apple case studies, 51–52, 55–56, 65, 68, 75–80, 86; in beef case studies, 187, 188, 194, 196–97, 198, 199, 208, *220*, 221; in blueberry case studies, 88, 94, 119–20, 123; in farm-direct channels, 296–98; in milk case studies, 241–42, 248, 257–59, 263n10, 263n11; in spring mix case studies, 138, 142, 144, 146–49, 156, 168, 170, *171*; and supply chain performance, 20–21; and supply chain size, 19–20

Roheim, Cathy A., 14

rural area funding programs, 335–37

Sacramento (CA): factors influencing prices in, 278, *284,* 287–88; farmers market pricing in, 280–82; food and agricultural industry in, 127–29; local production area in, 16; product availability in, 268–74

Sacramento Natural Foods Cooperative, 148–49

Safe Quality Food (SQF) program, 47

salads. *See* spring mix supply chains

sales. *See* revenues and sales

Salinas Valley (CA), 130. *See also* Earthbound Farm

San Benito County (CA), 130, *131*

Santa Monica Farmers Market, 318

scale economies. *See* economies of scale

Schmit, T., 279–80

seasonality. *See* product availability

Seniors Farmers Market Nutrition Program, 337

SFSC (Short Food Supply Chain) concept, 15

Shank, Edwin, 245. *See also* Shankstead EcoFarm

Shankstead EcoFarm: effect of regulations on, 253; food miles and fuel usage of, 249; growth potential of, 252–53; key lessons from, 253–54; local economic impacts of, 249–51; performance of, 251–52, 257–59; product flow chart of, *246*; and production, 245–47; structure of, 245

Short Food Supply Chain (SFSC) concept. *See* SFSC (Short Food Supply Chain)

short food supply chains. *See* direct market supply chains

size. *See* supply chain size

slaughtering. *See* processing systems

"Smart Choice Cafe" program, 69–70

SmartFresh ethylene technology, 65

Smith, Jim. *See* Jim Smith Farm

South Mountain Creamery: food miles and fuel usage of, 239–40, 260–61; impacts of regulations on, 242–43; key lessons from, 243–44; local economic impact of, 240–41; and marketing and distribution, 238–39; performance of, 241–42, 257–59; and processing and packaging, 237–38; product flow chart of, 237; and production, 236–37; structure and size of, 235–36, 255

Sowers, Karen, 236

Sowers, Randy, 236

spatially extended chains (SFSC), 15

spatial proximity chains (SFSC), 15

Spoonriver restaurant, 198–99

spring mix: factors influencing prices of, 284–85, 287; farmers market pricing of, 280–82; and price premiums for "local" attribute, 276–79; and product availability, 271–72; and production areas, 127–29, 130–31

spring mix supply chains: benefits of studying, 8–9; comparisons of performance in, 170–74; comparisons of structure in, 167–68; key lessons from, 174–75; local region defined, 127, 128; producers' net prices in, 297–98; and production and processing, 130–31. See also Davis Food Co-op; Earthbound; Fiddler's Green Farm; Nugget Markets

state food plans, 343

state food policies, 332

state meat processing inspection, 180–81, 225n5

Stephenson, G., 307n2

Stevenson, G. W., 21

Stewart, Carol, 183

Stewart, John, 183

storage: in apple case studies, 45–46, 54–55, 64–65, 67; in beef case studies, 185, 195, 210; in blueberry case studies, 91–92, 124n7; and product availability, 268, 269; in spring mix case studies, 136

St. Paul (MN). See Twin Cities (MN)

structure. See supply chain structure

SunShineHarvest Farm: community and economic impacts of, 200–201; comparison of performance of, 221–24; comparison of size and growth of, 219–21; comparison of structure of, 219; expansion prospects of, 201–2; food miles and fuel usage of, 199–200; key lessons from, 203–4, 224–25; and marketing and distribution, 195–99; and processing, 193–95; product flow chart of, 192; production, 191–93; structure and performance of, 190–91

SuperFoods (supermarket), 39–40; business relationships of, 74; comparisons of performance of, 76–81; comparisons of size and growth of, 75–76; and distribution and marketing, 43–48; fuel use of, 48–49, 80, 82; impacts of, 49, 79, 80–81; key lessons from, 51–52; and local apple program, 50–51; product flow chart for, 41; and product origin information, 73; product quality and safety standards of, 76, 324; suppliers of, 40–43

supermarkets. See Allfoods (supermarket); mainstream supply chains; SuperFoods (supermarket)

Supplemental Nutrition Assistance Program (SNAP), 56, 61, 334–35

supply and demand: balance between, 298, 320–21; as barrier to expansion, 19–20, 216, 234; demand forecasts, 44, 67; relation to pricing,

from, 217–18, 224–25; and Kowalski's Markets, 190; and marketing
and distribution, 210–12; and processing, 208–10; product flow chart
of, *205*; and production, 205–8;
retail sales for, 212–13; structure and
performance of, 204–5, *220*
traceability programs, 185–86, 315
transaction cost economics, 17
transportation. *See* distribution systems; food miles; fuel usage
Trickling Springs Creamery: affect of
regulations on, 253; food miles and
fuel usage in, 249, *260–61*, 299–300;
growth potential of, 252–53;
impacts of, 249–51; key lessons
from, 253–54; and marketing and
distribution, 247–49; performance
of, 257–59; and processing and
packaging, 247; product flow chart
of, *246*; structure of, 244–45, 255
Twin Cities (MN): farmers market
pricing in, 280–82; local production area defined for, 16; product
availability, 268–74; retail price
comparisons in, 283, 287–88; role of
food industry in, 178–80. *See also*
Kowalski's Markets; SunShineHarvest Farm; Thousand Hills Cattle
Company

Underhill, Paul, 155
United States Department of Agriculture (USDA): and the Census
of Agriculture, 7–8, 295–96; and
funding programs, 335–37; and the
GAP (Good Agricultural Practices) program, 47, 51, 96; and the

National Leafy Greens Marketing
Agreement, 140, 169

Valley Fabrication (manufacturer), 134
"variety" attribute, 56, 283, *286*
Verdoes, Alan, 184–85, 188
Veritable Vegetable, 154, *155*, 158–59,
162–63
Ver Ploeg, M., 303
vertical integration, 17, 39–40, 235
Vierra, David, 319
Vierra Farms, 319
Vogel, Stephen, 296–97

wages. *See* income and wages
Washington (state), 38
Washington (state) apple suppliers, 43,
44–45, 47, 48, 51–52
Washington DC: factors influencing
prices in, 278, *284*, 287–88; local
food availability in, 268–74; locality
defined for, 16, 228–29; milk indus-
try in, 230–31. *See also* Maryland
and Virginia Co-op; South Moun-
tain Creamery; Trickling Springs
Creamery
Waters, Alice, 129
Watson, J., 303
Weber, M. E., 300
Williamson, Oliver E., 17–18
willingness-to-pay studies, 275. *See
also* consumers
Wolf, M. M., 280
Women, Infants, and Children (WIC)
Program, 56, 337, 339

year-round availability. *See* product
availability

To order or obtain more information on these or other University of Nebraska Press titles, visit www.nebraskapress.unl.edu.